AQA
Biology
for A2

■ Mike Bailey
■ Bill Indge
■ Martin Rowland

DYNAMIC LEARNING
Innovate • Motivate • Personalise

HODDER
EDUCATION
AN HACHETTE UK COMPANY

The Publishers would like to thank the following for permission to reproduce copyright photos:

p.iv *l* Mark Hamblin/agefotostock/Photolibrary.com, *r* Bill Indge; **p.1** Mike Powles/Oxford Scientific (OSF)/Photolibrary.com; **pp3, 4, 7** Bill Indge; **p.12** DEA/P. Jaccod/Photolibrary.com; **pp13, 18, 25, 31** *both* Bill Indge; **p.32** Keith Ringland/Oxford Scientific (OSF)/Photolibrary.com; **pp35, 37** Bill Indge; **p.43** Dr Jeremy Burgess/Science Photo Library; **p.48** Phil Degginger/Alamy; **pp50, 51, 55, 58, 59** *both* Bill Indge; **p.64** Harold Taylor/Oxford Scientific (OSF)/Photolibrary.com; **p.65** *t* Mary Beth Angelo/Science Photo Library, *c*, *b* Bill Indge; **p.70** Bernd Zoller/imagebroker.net/Photolibrary.com; **pp73, 75** Bill Indge; **p.76** Mike Grandmaison/Photolibrary.com; **pp80, 89** *both*, **94** Bill Indge; **p.99** N. A. Callow/NHPA; **p.102** Grant Heilman Photography/Alamy; **p.106** Wisconsin Fast Plants Program, University of Wisconsin – Madison; **p.119** Ernie Janes/NHPA; **p.120** *t* superclic/Alamy, *b* Mark Boulton/Alamy; **p.124** Dave Watts/Alamy; **p.126** Eye of Science/Science Photo Library; **p.127** Phototake Science/Photolibrary.com; **p.128** *t* a) Alan & Sandy Carey/Cusp/Photolibrary.com, b) Martin Harvey/NHPA, c) Yves Lanceau/NHPA, d) Arco Images GmbH/Alamy, e) Image Register 044/Alamy, *b* Steve Hopkin/Ardea; **p.129** Peter Falkner/Science Photo Library; **p.130** Maria Taglienti-Molinari/Brand X/Corbis; **p.132** Juniors Bildarchiv/Photolibrary.com; **p.135** *t* Eric B. Guinther, AECOS, *b* Lester V. Bergman/Corbis; **p.139** *t* Maximilian Stock Ltd/Photolibrary.com, *b* Graham Jordan/Science Photo Library; **p.142** Image Source Black/Alamy; **p.144** CNRI/Science Photo Library; **p.150** Institut Kage/Photolibrary.com; **p.151** Warren Anatomical Museum, Francis A. Countway Library of Medicine; **p.153** Science Pictures Ltd/Science Photo Library; **p.162** Marevision/agefotostock/Photolibrary.com; **p.163** Dr Don Fawcett/Visuals Unlimited/Getty Images; **p.165** Manfred Kage/Science Photo Library; **p.171** Kristen Anderson; **p.172** *l* Winfried Wisniewski/agefotostock/Photolibrary.com, *r* Dan Hartman/NHPA; **p.174** *l* Lehtikuva Oy/Rex Features, *r* Mark Dadswell/Getty Images; **p.177** *l* Steve Gschmeissner/Science Photo Library, *r* Biology Media/Science Photo Library; **p.179** Don Fawcett/Science Photo Library; **p.186** Georgie Holland/agefotostock/Photolibrary.com; **p.189** *t* Don Fawcett/Science Photo Library, *b* Scott Camazine/Science Photo Library; **p.194** Nigel Cattlin/FLPA; **p.195** Al Fenn/Time Life Pictures/Getty Images; **p.196** Harry Walker/Index Stock Imagery/Photolibrary.com; **p.198** Edward Kinsman/Science Photo Library; **p.200** Sebastien Boisse/Photononstop/Photolibrary.com; **p.201** Daniel Heuclin/NHPA; **p.202** Erwin & Peggy Bauer/Photolibrary.com; **p.205** Collection CNRI/Phototake Science/Photolibrary.com; **p.213** *l* Richard Becker/FLPA, *c & r* Emanuele Biggi/Oxford Scientific (OSF)/Photolibrary.com; **p.214** *tl* London Scientific Films/Oxford Scientific (OSF)/Photolibrary.com, *tr* blickwinkel/Alamy, *c* Joe Blossom/NHPA; **p.215** Najlah Feanny/Corbis SABA; **p.226** *tl* J. Testart/ARFIV/Science Photo Library, *tc* Eric Grave/Phototake Science/Photolibrary.com, *tr* Prof. S. Cinti/Science Photo Library, *bl* Institut Kage/Doc-Stock/Photolibrary.com, *bc* Science Photo Library/Photolibrary.com; **p.227** *t* Peter Menzel/Science Photo Library, *c* Adrian Davies/Alamy; **p.231** John Davidson/Alamy; **p.234** Javier Larrea/agefotostock/Photolibrary.com; **p.248** Jayanta Shaw/Reuters/Corbis; **p.250** *c* Maximilian Stock Ltd/Science Photo Library, *l* Grant Heilman Photography/Alamy; **p.251** Nick Cobbing/Rex Features

t = top, *b* = bottom, *l* = left, *c* = centre, *r* = right

The Publishers also acknowledge the following sources of data:

p.158 Table 1 data from *Review of Medical Physiology*, W. F. Ganong, Prentice Hall, 1995; **p.176** Table 1 data from *Exercise Physiology*, W. McArdle *et al.*, Lippicott, Williams & Wilkins, 2001

Every effort has been made to trace all copyright holders, but if any have been inadvertently overlooked the Publishers will be pleased to make the necessary arrangements at the first opportunity.

Although every effort has been made to ensure that website addresses are correct at time of going to press, Hodder Education cannot be held responsible for the content of any website mentioned in this book. It is sometimes possible to find a relocated web page by typing in the address of the home page for a website in the URL window of your browser.

Hachette UK's policy is to use papers that are natural, renewable and recyclable products and made from wood grown in sustainable forests. The logging and manufacturing processes are expected to conform to the environmental regulations of the country of origin.

Orders: please contact Bookpoint Ltd, 130 Milton Park, Abingdon, Oxon OX14 4SB. Telephone: (44) 01235 827720. Fax: (44) 01235 400454. Lines are open 9.00 – 5.00, Monday to Saturday, with a 24-hour message answering service. Visit our website at www.hoddereducation.co.uk

© Mike Bailey, Bill Indge, Martin Rowland 2009
First published in 2009 by
Hodder Education, an Hachette UK Company
338 Euston Road
London NW1 3BH

Impression number	5 4 3
Year	2013 2012 2011

Cover photo: SCIENCE PHOTO LIBRARY
Illustrations by Barking Dog Art
Typeset in Palatino 10pt by Fakenham Prepress Solutions, Fakenham. Norfolk NR21 8NN
Printed in Dubai

A catalogue record for this title is available from the British Library

ISBN: 978 0340 94619 0

Contents

Introduction

Figure 1 is a photograph of a grey seal. This animal is a marine mammal and it is quite common along parts of the British coast. Grey seals are true seals and belong to the family Phocidae. Fur seals, such as the animal shown in Figure 2, belong to a different family, the sea lions or Otariidae. Members of these two families are similar in appearance and it is thought that, a long time ago, they shared a common ancestor. In this book we shall be looking at the genetic variation that exists between organisms such as seals and how, over time, different species, genera and families have evolved.

▲ **Figure 1** A grey seal living off the North Lincolnshire coast. Grey seals are true seals and belong to the family Phocidae.

▲ **Figure 2** An Australian fur seal. Fur seals belong to the sea lion family, Otariidae.

Seal populations are influenced by human activity. Seals, for example, compete with humans for fish stocks. The topic of ecology forms an important theme in your A2 Biology course. We shall be studying the features of communities and ecosystems in order to understand how the sustainability of resources depends on managing the conflict between human need and conservation.

Seals are remarkable divers. An elephant seal, for example, can remain under water for up to 80 minutes before coming to the surface to breathe. Seals have adaptations that allow them to remain below the surface of the cold waters of the Arctic and Antarctic seas for long periods. They have insulating layers of blubber or fur; their muscles respire anaerobically and tolerate high concentrations of lactate and they have adaptations that allow them to conserve oxygen. We shall be looking at some of the ways in which animals such as seals are able to detect stimuli and respond to them. In this way they increase the probability of their survival by moving away from harmful environments and maintaining optimal internal conditions for their metabolism.

Biologists who study seals and other marine mammals encounter particular difficulties because the animals they are studying often breed in wild and isolated places and spend much of their time beneath the surface of the sea. Knowledge of DNA has not only resulted in a better understanding of the evolutionary relationships between seals but has also allowed biologists to recognise individual animals and study movements and migrations. This is an example of the final theme explored in your A2 course; the way in which DNA controls the metabolic activities of cells and some of the medical and technological applications that have resulted from this knowledge.

So, if you want a broad picture of the biology of seals, you will need to study their genetics, their ecology and their physiology, and understand how all these different aspects are linked together. Although this book is divided into chapters and each chapter concentrates on a particular topic, it is important to appreciate how all these areas of biology are related to each other and how much they rely on the basic biological principles covered in your AS course.

About this book

This book has been written in a similar way to the AS textbook, *AQA Biology for AS*. Each chapter shares a number of features.

The chapter opening

We have chosen topics that relate to a broad theme that underlies each chapter to introduce the chapters in the book. Many of these topics illustrate how scientists use their knowledge and understanding to ask questions and suggest hypotheses, how they carry out experimental work and interpret and evaluate their results, and how they make use of their results in ways that may benefit humans.

The text

Your AS course should have provided you with a sound understanding of many fundamental biological ideas. The material in this book builds on these concepts and develops many of them further. In writing each chapter, we have aimed at providing you with a good understanding of basic principles rather than with a lot of unnecessary detail and too much technical language. The diagrams, photographs and information contained in the captions should provide you with additional help in gaining an understanding of the subject.

Progress questions

Progress questions are all in blue boxes and are identified with the letter 'Q'. They should help you to understand what you have been reading and are meant to be answered as you go along. Some of them are very straightforward and can be attempted from information in the paragraphs which come immediately before. Others want you to apply something that was explained earlier to a new situation. Your teacher has access to the answers.

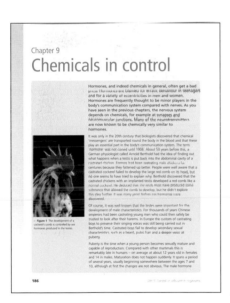

Q12

The population of some countries is decreasing. Describe the main features of an age pyramid representing a decreasing population.

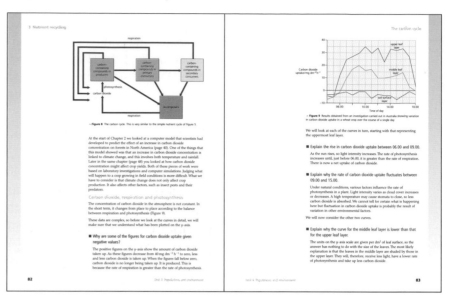

The questions in these sections are in blue and have square bullet points in front of them. These exercises provide opportunities for you to develop the skills of application and data analysis. They are based on photographs and diagrams, graphs and tables and take you through the exercise, one step at a time. To get the most out of them, you should try to answer the questions *before* you look at the answers that are given underneath.

Interpretation exercises

These exercises have a yellow background. They will provide you with an opportunity to work through more complex exercises on your own.

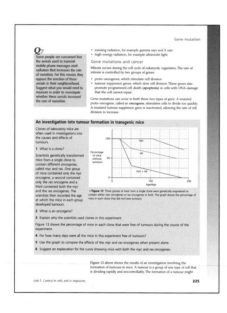

Link to web-based material

This book is supported by a free website at:

www.dynamic-learning-student.co.uk

Once you have registered you can access resources such as:

- learning outcomes
- photographs
- personal tutorials
- web links.

Details of how to register are given on the inside front cover.

This icon indicates that there is a related item on the website.

Chapter 1
Populations

The tiger is one of the world's rarest wild mammals. Poaching for skins and for body parts used in traditional Chinese medicine, and conflict with increasing human populations have led to a huge decrease in the number of wild tigers. Biologists are trying to protect the remaining tiger populations. They must monitor numbers as accurately as possible. We will look at a study carried out in the Way Kambas National Park on the Indonesian island of Sumatra. Figure 1 shows a map of the park.

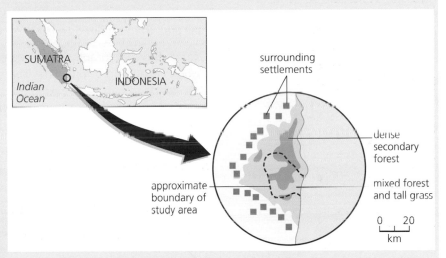

▲ **Figure 1** The Way Kambas National Park.

▲ **Figure 2** Every tiger has a unique pattern of stripes. The scientists analysed the pictures taken by the cameras and were able to identify individual animals from their stripe pattern.

The map shows that the national park is surrounded by human settlements. People from these settlements graze their cattle in the park grasslands and cut wood from the park's forests for fuel for their cooking fires. The tigers living in the park may become extinct before long. Scientists trying to protect these tigers needed an accurate count of the number of animals but this proved very difficult. The map shows that the park vegetation is a mixture of dense forest and tall grass and it is very difficult to find tigers in these conditions. The scientists used technology to help them to find out the number of tigers in a study area within the park. They set up automatic cameras at selected points along trails frequented by tigers. Each time an animal passed through an infra-red beam crossing the trail, the camera took a photograph (Figure 2).

The graph in Figure 3 shows some of the results from this investigation.

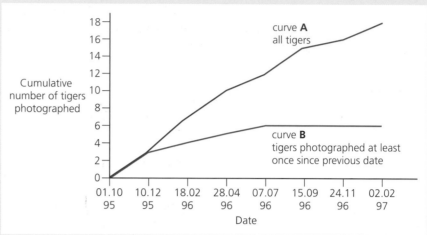

▲ **Figure 3** Graph showing numbers of tigers photographed in Way Kambas National Park study area between 1995 and 1997.

Tigers are territorial animals. A tiger spends most of its time in a relatively small area that it defends against other tigers.

- Curve **B** on the graph shows tigers that were photographed at least once since the previous date. This suggests that these tigers were resident and had territories in the study area. You will see that this curve rises at first then levels out at a value of 6. The scientists were able to conclude that six tigers had territories in the study area.

- Curve **A** represents all the tigers that were photographed. It includes the tigers that had territories as well as animals that just passed through the study area. You will see that this curve does not level out.

Scientists must be very careful in interpreting data such as those collected from this study. Curve **A** suggests that there were at least 18 tigers in the study area. This figure obviously includes tigers that live outside the area and occasionally wander into it. Curve **B**, however, gives a much better estimate of the tiger population in the study area. It shows just how low this population has become.

If we want to understand the ecology of a particular area we first need to get some idea of the number of organisms of different species that are found there and how they are distributed. We can then begin to use our understanding of ecological principles to explain these observations. In this chapter we will start by looking at different methods ecologists use to find out about numbers and distributions. We will then consider some of the factors that influence population size before going on to consider some aspects of human populations.

Organisms and their environment

Nettles are common British plants. You can find them growing in woods, on grazing land, by the sides of ponds and streams, and in areas of wasteland in towns and cities. Figure 4 shows a clump of nettles and some of the animals that can be found on them.

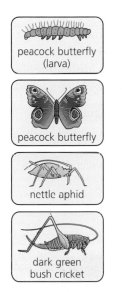

▲ **Figure 4** Some common insects associated with a clump of nettles.

The set of conditions that surrounds the insects in this clump of nettles is called the **environment**. It consists of abiotic factors and biotic factors.

- **Abiotic factors** make up the non-living part of the environment. Nettles grow well, for example, where there is a high concentration of phosphate ions in the soil. Warm, humid conditions result in large numbers of aphids on the leaves. The concentration of phosphate ions, temperature and humidity are all abiotic factors.
- **Biotic factors** are those relating to the living part of the environment. The numbers of aphids on the nettles will be affected by the numbers of ladybirds because ladybirds feed on aphids. Predation by ladybirds is a biotic factor.

A **population** is a group of organisms belonging to the same species. The members of a population are able to breed with each other so the two-spot ladybird population consists of all the two-spot ladybirds found in a particular area at a particular time. A **community** is the term used to describe all the populations of different organisms living in the same place at the same time. When we talk about the community in this nettle patch, we mean the nettles, the aphids, the ladybirds and all the other organisms that we have not mentioned, such as the fungi and the bacteria that live in the soil surrounding the roots of the nettles.

There are two more terms that we often use to explain ecological ideas. The **habitat** of an organism is the place where it lives. Its **niche** describes not only where it is found but what it does there. We can describe the niche of the two-spot ladybird, for example, in terms of the abiotic features of its habitat, such as the temperature range it can tolerate, and the position on the nettle plant where it is found. We can also bring into our description some idea of its feeding habits, referring to the size and species of aphids that it eats.

Counting and estimating

Every ten years there is a census of all the people living in the UK. It is not totally accurate because some people fail to fill in their census forms accurately, or don't fill them in at all, but it does give a good idea of the size of the UK human population.

There are other species where we can get a fairly accurate figure for the population by counting them. For example, seabirds nesting on the Farne Islands off the north-east coast of England can be counted because they nest in large colonies – but the data obtained are much less accurate than for human populations.

▲ **Figure 5** Many species of seabird, such as these guillemots, nest in dense colonies. Biologists monitor seabird populations by counting the number of breeding birds.

We can only count all the animals or plants in a population in a few cases. The organism concerned needs to be large, conspicuous and confined to a relatively small area for a count to be accurate. If it is not possible to count every single organism, we need to take samples. Ecologists usually base their estimates of populations on samples. They must make certain, however, that the samples are representative of the population as a whole. In order to be sure of this, samples must be large enough to be representative and taken at random.

Sample size

The larger the size of a sample, the more reliable is the result. Data from a very large sample, however, cannot be collected accurately in the time available. When we sample an area we have to strike a balance. Look at the graph in Figure 6. It shows that a very small sample has very few species in it. As the sample size increases, the number of species that it contains increases. There comes a time when the number of species does not

Q1

Suggest reasons why estimates of seabird numbers based on counting breeding birds may not be accurate.

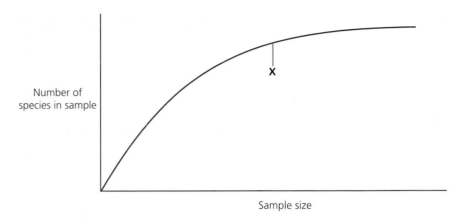

▲ **Figure 6** The effect of sample size on the number of species present. After point **X**, the number of species does not increase much more with increasing sample size. The sample size at **X** is therefore representative of the community as a whole. A larger sample would simply mean more work.

increase much more however much bigger the sample. This sample size is obviously representative of the community as a whole.

Random sampling

When you studied the topic of variation in Unit 2 you learnt that, unless samples were taken at random, the results may be biased. The same is true of ecological samples. They must be collected at random or, again, the results may be biased. Quadrats are often used to mark out areas to be sampled. When we use a quadrat for sampling, it is important that we use a method that will result in the quadrat being genuinely placed at random in the study area (Figure 7).

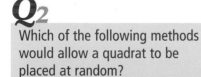

Q2

Which of the following methods would allow a quadrat to be placed at random?

A Closing your eyes, turning on the spot and throwing the quadrat over your shoulder.

B Placing quadrats along a tape at 5 m intervals.

C Picking random numbers from a hat to give co-ordinates on a grid.

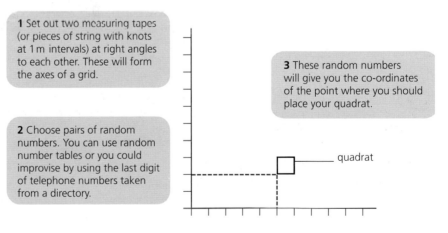

1 Set out two measuring tapes (or pieces of string with knots at 1 m intervals) at right angles to each other. These will form the axes of a grid.

2 Choose pairs of random numbers. You can use random number tables or you could improvise by using the last digit of telephone numbers taken from a directory.

3 These random numbers will give you the co-ordinates of the point where you should place your quadrat.

quadrat

▲ **Figure 7** How to place a quadrat randomly.

Quadrats and transects

Two important ways of sampling involve quadrats and transects. Both of these are in general used for plants, but they can also be used for other organisms that do not move about very much, such as many of those that

1 Populations

Sampling method	Method	What it is used for	How it is used
Quadrat	A defined area in which to sample organisms	Studying the distribution of plants or animals that do not move about much in a fairly uniform area	Placed at random and used to find population density, frequency or percentage cover. In some investigations, such as those involving the effects of grazing, permanent quadrats may be used. They may remain in place for many years.
Transect	A line through a study area along which samples are taken	Usually used where the environment and species gradually vary	Placed so that it follows the environmental gradient, for example, up a seashore or from sunlight into shade. Quadrats may be used at regular intervals along a transect to sample in more detail.

Table 1 Using quadrats and transects.

live on seashores. Table 1 shows when and how these techniques are commonly used. There are three different ways that we use to describe the distribution of organisms once the quadrat or transect is put in position. These are illustrated in Figure 8.

a) Population density

This quadrat measures 0.5 m × 0.5 m. It contains six dandelion plants. The **population density** of dandelions would be 24 plants per m^2. To get an accurate figure you would need to collect the results from a large number of quadrats.

b) Frequency

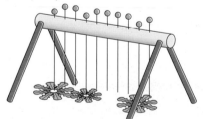

This point quadrat frame is being used to measure **frequency**. The pins of the frame are lowered. Suppose three out of ten pins hit a dandelion plant. The frequency of dandelion plants will be three out of ten, or 30%.

c) Percentage cover

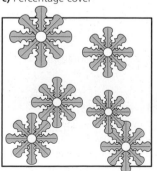

Percentage cover measures the proportion of the ground in a quadrat occupied by a particular species. The percentage cover of the dandelions in this quadrat is approximately 40%.

▶ **Figure 8** Using **a)** population density, **b)** frequency and **c)** percentage cover to describe the distribution of dandelion plants.

Mark-release-recapture

Most animals move around so it is not usually possible to estimate the size of an animal population using a quadrat. Instead we use the mark-release-recapture method. This relies on capturing a number of animals and harmlessly marking them so that they can be recognised again. The marked animals are then released. Some time later a second sample is trapped and the numbers of marked and unmarked animals in the second sample are recorded. From the data collected, the size of the population can be estimated. The calculation relies on the fact that the proportion of marked animals in a sample (the second sample) is the same as the proportion of marked animals in the whole population. Here is an example of a calculation used to estimate the size of a population of grasshoppers:

number of grasshoppers caught, marked and released	= 56
number of marked grasshoppers in second sample	= 16
total number of grasshoppers in second sample	= 48

proportion of marked grasshoppers in sample = proportion of marked grasshoppers in population

$$\frac{\text{number of marked grasshoppers in sample}}{\text{total number of grasshoppers in sample}} = \frac{\text{number of marked grasshoppers in population}}{\text{total number of grasshoppers in population}}$$

$$\frac{16}{48} = \frac{56}{\text{total number}}$$

Rearranging this equation:

$$\text{total number} = \frac{48 \times 56}{16}$$

$$= 168$$

Scientists have modified the mark-release-recapture technique to estimate the size of whale populations.

Q3

A sample of 40 trout in a fish pond was caught in a net. Each fish was harmlessly marked on one of its fins. The fish were then released back into the pond. One week later a second sample was caught. It contained 17 marked trout and 41 unmarked trout. Estimate the total number of trout in the pond.

▲ **Figure 9** Because whales spend most of their time below the surface of the sea they cannot be trapped or easily marked, so scientists needed to find a method other than observing and counting the whales in order to estimate the size of whale populations. They developed a DNA fingerprinting method.

Q4

Use the information in the list on the right to suggest how scientists could modify the mark-release-recapture technique to estimate the size of a whale population.

- Whales cannot be trapped and they cannot easily be marked.
- Scientists can use a small boat to approach a surfaced whale. They can remove a tiny piece of skin from the whale.
- The scientists are able to analyse the DNA from this piece of skin and produce a DNA fingerprint. Each whale has its own unique DNA fingerprint.

A population rarely remains constant. Numbers of a particular species vary from place to place and from one time of year to another. These variations may result from differences in **abiotic** factors. They may also result from changes in **biotic factors** (see page 15).

Chance, probability and population size

Look at a roadside verge in early spring and you will almost certainly see dandelions in flower. Examine a particular length of verge more closely and you will see that most of the dandelions are growing near the road. Further away they are less common and other species are more numerous. How can we explain this observation? Perhaps the road was treated with salt during the winter and dandelions can tolerate high salt concentrations better than other plants. There is another explanation, however, that has very little to do with biology. Maybe it is due to chance. It could be that we just picked an area where there were more dandelions growing closer to the road.

We will look at another investigation in more detail. Observations suggest that birds are able to distinguish different colours:

- Insect-eating birds learn to avoid distasteful species such as ladybirds. Many ladybirds are bright red.
- The berries of many plants turn red as they ripen. Blackbirds eat ripe red hawthorn berries but they do not eat unripe green ones.
- Slug pellets, which gardeners use to kill slugs and snails, are coloured blue. The manufacturers claim that birds avoid blue, so won't be poisoned by feeding on slug pellets.

A group of students decided to investigate this further and tested colour choice by blackbirds. They dyed oatmeal one of four different colours: red, green, yellow or blue. The dyed oatmeal was dipped in fat and put out on a tray. The students kept records of the birds visiting the tray and the colours of the food they ate. Each week the results of 50 visits were recorded. Some of the results from the investigation are shown in Table 2 opposite.

What conclusions could the students draw from these data? Assuming that the behaviour of blackbirds in this investigation is linked to their feeding behaviour with their normal food, it would seem fair to suggest that they had a distinct preference for red and that they seldom selected blue. We would have little hesitation in coming to this conclusion.

Now look at the difference between blackbirds feeding on yellow-coloured food and those feeding on green-coloured food. It is not easy to arrive at a firm conclusion from the evidence in the table. Although the mean value for those eating yellow food was higher than that for those eating green food, the difference is not large and, in addition, there was considerable

Week	Number of times blackbirds fed on food coloured:			
	red	green	yellow	blue
1	32	13	5	0
2	38	3	7	2
3	29	6	14	1
4	36	2	12	0
Mean	**33.8**	**6.0**	**9.5**	**0.8**

Table 2 The effect of colour on food eaten by blackbirds.

variation from one week to the next. The difference between the results may simply be down to chance. We need a more precise way of looking at our results than relying on personal opinion. We must apply an appropriate **statistical test**. This allows us to calculate the probability of obtaining results that differ from each other by a particular amount purely by chance.

In this particular case, carrying out a suitable statistical test allows us to make the statement that the probability of blackbirds selecting yellow rather than green food is approximately 19 out of 20. We could turn this statement round and say that there is only a probability of 1 in 20 that these results arise by chance.

We accept different levels of probability in different situations. Suppose you were offered a 10p raffle ticket in a draw for which the first prize was a car. You would probably be happy to buy the ticket if there was a 1 in 20 probability of winning the car! You might even be very tempted to buy a ticket if the probability of winning the car was 1 in 100, or even 1 in 1000. Now look at another situation. Suppose investigators established a link between eating baked beans and developing cancer. You would probably be very reluctant to accept a risk of 1 in 1000 and baked beans would rapidly disappear from your diet.

As biologists, we understand that we can never completely rule out the effect of chance. Look again at Table 2. We felt quite confident in saying that on the evidence in the table blackbirds preferred red food to green food, but there is still a small probability that the results we obtained were due to chance. In fact this probability is very low indeed, a lot less than 1 in 1000. It seems quite reasonable to say that the probability of results like these arising due to chance is so low that we can safely reject chance as an explanation. We need a cut-off point and, for biological investigations, we normally accept this as a probability of 1 in 20 or 0.05. If there is a probability of less than 1 in 20 that our results could have arisen by chance, then we recognise that there must be another explanation.

Different statistical tests for different purposes

A statistical test is a tool and, like all tools, you don't need to know how it works to make use of it. If you have an electric drill or a food processor, it is not necessary to understand how an electric motor works in order to use it.

What you need to be able to do, however, is to select the right tool for the right job.

The investigations that you are likely to carry out during your A-level Biology course generate data that you may need to look at in different ways. It is likely that you will want to do one of the following.

- **See if there is a difference between two or more sets of data** – Suppose you were investigating the effect of temperature on the growth of duckweed. You might want to see if there was a significant difference between the numbers of duckweed plants when the plants were kept at two different temperatures.
- **See if there is an association between two sets of data** – If you investigated the number of sea aster plants growing at different distances from a tidal creek, you might want to see if there was a significant association between the number of sea aster plants and the salinity of the soil.
- **Investigate data that fall into distinct categories** – This is the sort of situation that often arises in investigations involving behaviour. Suppose you were looking at whether woodlice moved to light or dark conditions. You may want to know if there is a significant difference between the data you have collected and the data which your hypothesis predicts.

Each of these situations requires a different statistical test. The decision chart in Figure 10 should help you to choose which one to use.

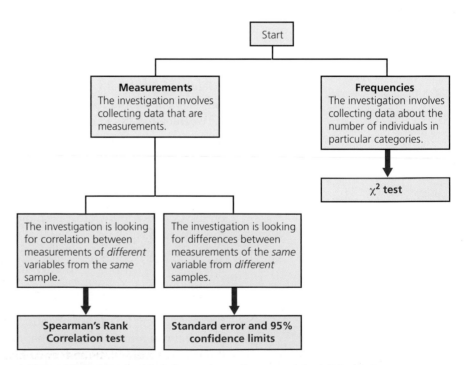

▲ **Figure 10** Decision chart to help you choose the appropriate statistical test.

Making the decision over which test to use is very important. The statistical test you use and the way you design your investigation are closely linked. If you want to put up a shelf, the decision to use either screws or nails will determine whether you need a screwdriver or a hammer. In the same way, it is essential when you design an investigation to bear in mind the statistical test you might want to use. In this way you can be sure that you will collect sufficient data in a form that can be tested.

Carrying out statistical tests

The null hypothesis

All of the statistical tests that you will use in your A-level Biology course are based on what is called a **null hypothesis**. It is very important, therefore, that you understand what a null hypothesis is and why it is used. We will look at an example. Suppose you were investigating the distribution of aphids on nettle leaves. You collected samples from the upper surface and from the lower surface. You need to have a clear hypothesis that you can test. It is difficult to predict what you will find here – after all, this is an investigation, so you probably don't know. It is much better to word our hypothesis in terms of what will happen if there is no difference. This is why we call it a null hypothesis. In this example, our null hypothesis could be:

There is no difference between the numbers of aphids on the upper surface and on the lower surface of nettle leaves.

You can then carry out your statistical test and, as a result, come to a conclusion about whether or not to accept the null hypothesis.

- If the test indicates that you should accept the null hypothesis – then all that you can say is that there is no significant difference between the results on the different surfaces and that the data you collected could be explained purely by chance.
- If the test indicates that you should reject the null hypothesis – then you can say that the results are unlikely to be due to chance and you can accept that there is a biological explanation.

Calculating the test statistic

Once you have set up your null hypothesis, you can carry out the test you have selected. As a biologist, carrying out statistical tests is like using a recipe book to cook a meal. Find an example of the test that does what you want, then amend it by substituting your data for those in the example. If you work methodically, and can use a standard scientific calculator, there is nothing to worry about. It is all very straightforward. There are many different statistical tests but, if you look on the pages listed below, you will find worked examples of the tests you require.

- Standard error and 95% confidence on page 12.
- Chi-squared (χ^2) test on page 22.
- Spearman's rank correlation test on page 94.

Interpreting the results

Calculating the test statistic gives you a value. What does this value mean? You need to interpret this figure in terms of probability. You can then make a decision as to whether to accept or reject your null hypothesis. In order to do this, you may need to look up the value that

▲ **Figure 11** The individual common duckweed plants shown here are about 5 mm across.

you have calculated in a set of statistical tables where the probability has been calculated for you.

How abiotic factors can affect a population

Investigating the effect of temperature on duckweed

Common duckweed (Figure 11) is a tiny plant that floats on the surface of the water in ponds and streams. It grows very rapidly, so it is a very useful organism to use for investigating the effect of abiotic factors, such as temperature or pH, on growth rate. Scientists investigated the effect of temperature on the growth of duckweed. They put 20 duckweed plants into a number of shallow dishes of water. They kept these dishes at different temperatures. Six days later they counted the number of plants in each dish. Some of their results are shown in Table 3.

Number of duckweed plants in dish after six days											
Dishes kept at 12 °C						Dishes kept at 22 °C					
37	38	38	36	36	43	60	56	54	60	58	62
40	32	37	39	38	36	62	61	60	63	59	53
40	38	38	39	34	39	58	57	59	58	61	57
39	38	34	37	38	37	61	59	61	60	56	62
35	39	38	41	37	39	64	53	58	61	60	59

Table 3 The effect of temperature on the growth of duckweed plants.

We use standard error and 95% confidence limits to find out if there is a significant difference between two means, as follows.

- **Produce a null hypothesis.** There is no difference between the number of duckweed plants in the dishes kept at 12 °C and the number of plants in the dishes kept at 22 °C.
- **Calculate the standard error.** Use your calculator to work out the mean and standard deviation of each of the samples (Table 4).

	Dishes at 12 °C	Dishes at 22 °C
Mean	37.7	59.1
Standard deviation	2.2	2.7

Table 4 Mean and standard deviation for duckweed plants.

Calculate the standard error of the mean, *SE*, for each sample using the following formula:

$$SE = \frac{SD}{\sqrt{n}}$$

where *SD* = standard deviation; *n* = sample size

Table 5 shows these values.

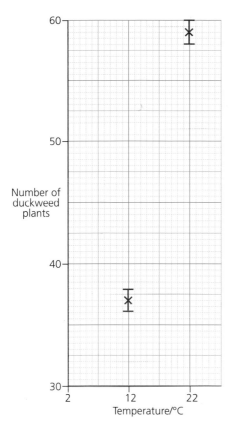

	Dishes at 12 °C	Dishes at 22 °C
Mean	37.7	59.1
Standard deviation	2.2	2.7
Standard error	0.4	0.5

Table 5 Mean, standard deviation and standard error for duckweed plants.

▲ **Figure 12** Using a graph to interpret standard error.

The 95% confidence limits are 1.96 standard errors above the mean and 1.96 standard errors below the mean (1.96 can be rounded up to 2). If the 95% confidence limits do not overlap, there is a 95% probability that the two means are different. In other words, there is a significant difference between the means at the $p = 0.05$ level of probability, and the null hypothesis should be rejected.

Figure 12 shows this information plotted as a graph. We have plotted the mean values for each sample as a cross and drawn bars to represent 2 standard errors on either side of these mean values. In this case, there is no overlap between the bars. We can therefore say that there is a probability of less than 0.05 that these results occurred by chance. We reject the null hypothesis and say that there is a significant difference between the results obtained at 12 °C and those obtained at 22 °C.

Lichens and pollution

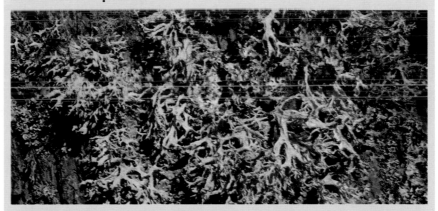

▲ **Figure 13** Lichens growing on a tree trunk.

A lichen (Figure 13) is made up of two different organisms. It is a partnership between an alga and a fungus. If you examine a section through a lichen with a microscope you will see green algal cells among thread-like fungal hyphae.

Sulphur dioxide is a major air pollutant which can reach high concentrations in industrial areas. Lichens are very sensitive to sulphur dioxide. Scientists have shown that the rate of carbon dioxide uptake in lichens is affected by atmospheric sulphur dioxide concentration. Table 6 shows the effect of sulphur dioxide concentration on the rate of uptake of carbon dioxide by three different species of lichen.

Mean sulphur dioxide concentration/ arbitrary units	Rate of uptake of carbon dioxide by lichen:		
	Usnea subfloridana	*Hypogymnia physodes*	*Leconora conizaeoides*
0.0	300	330	280
0.1	320	440	n/a
0.2	210	340	320
0.3	30	390	n/a
0.4	0	0	280
0.6	0	0	40
0.8	0	0	10

n/a: Data not available

Table 6 The effect of sulphur dioxide concentration on the rate of carbon dioxide uptake by three species of lichen.

1 There are no units in this table for the rate of uptake of carbon dioxide. Suggest units that could be used so that the rates of uptake can be compared reliably.

2 Suggest why *Usnea subfloridana* cannot survive when the mean sulphur dioxide concentration is greater than 0.3 arbitrary units.

3 Which of the three lichens would you be most likely to find growing in a city centre? Explain your answer.

4 Ecologists have suggested that it is more useful to monitor sulphur dioxide concentration by finding the diversity of lichens than by recording the presence or absence of particular species. Use the data in the table to suggest why.

A team of Italian scientists developed a method of using the diversity of lichens to monitor sulphur dioxide concentration. The method below is based on the protocol they developed.

- Choose five trees with straight trunks and with circumferences greater than 70 cm. Avoid trees that are leaning over or are covered in moss.

- Take a sampling grid measuring 30 × 50 cm divided into rectangles each measuring 15 × 10 cm. Position this on the same side of each tree at a height of 1.5 m above the ground.

- For each species of lichen, record the number of rectangles in which it is found. This will give its frequency on a scale of 1 to 10.

- Calculate the mean diversity index for the five trees in the sample by adding the frequencies of all the individual species and dividing by five.

- Repeat this to show the pattern of lichen diversity in a particular area.

5 Suggest why the trees to be sampled should have trunks greater than 70 cm in circumference.

6 Trees that lean should not be sampled. Suggest why.

7 a) Suppose all the tree trunks in one particular area were covered with a single species of lichen. What would be the mean diversity index for these trees?

b) How would you expect the mean diversity index to change if there were more lichens present?

Look at Figure 14 which shows some results from the Italian scientists' investigation.

8 Suggest an explanation for the similar patterns in the two maps.

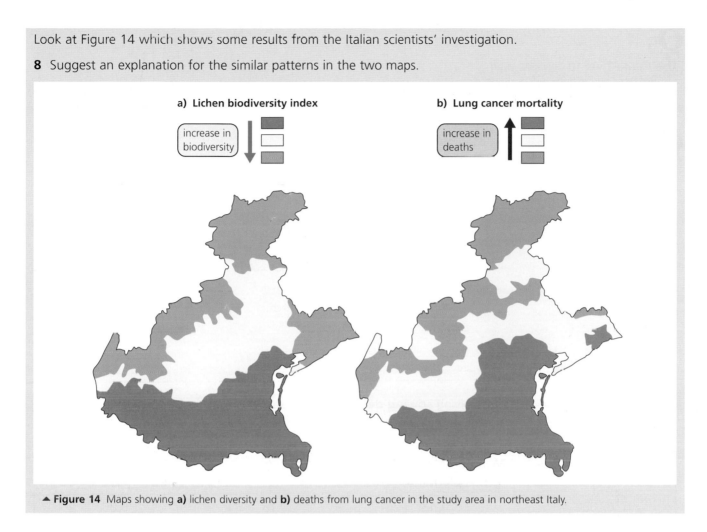

a) Lichen biodiversity index

increase in biodiversity

b) Lung cancer mortality

increase in deaths

▲ **Figure 14** Maps showing **a)** lichen diversity and **b)** deaths from lung cancer in the study area in northeast Italy.

Competition

Farmers grow crops for profit. One of the factors they take into consideration is the number of seeds they sow. The number of seeds sown per unit area is called the **sowing density**. Agricultural scientists have carried out research on how sowing density affects crop yield and weed growth. In one investigation they sowed wheat at different densities and measured the mass of grain produced by each plant and the mass of weeds. Some of their results are shown in the graphs in Figure 15, overleaf.

Look first at Figure 15 a). You can see that at sowing densities greater than 100 seeds m^{-2} the grain yield decreased with increasing sowing density. This can be explained in terms of **intraspecific competition** – competition between organisms that belong to the *same* species. The more wheat plants there are per square metre, the more competition there will be for resources such as light, water and mineral ions. The resources available to each plant will be fewer. This will result in smaller plants that produce less grain.

Now look at Figure 15 b). This is an example of **interspecific competition** – competition between *different* species. In this case, the wheat plants are competing with the weeds. The higher the density of wheat plants, the fewer the resources there are available for the weeds.

1 Populations

Q5

Use Figure 15 a) to describe how sowing density affects grain yield in tonnes per hectare.

Q6

A farmer could use the results of this investigation to find the optimum sowing density for wheat. What conclusion do you think he would draw? Explain your answer.

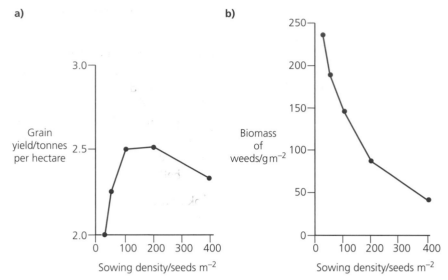

▲ **Figure 15** Graphs showing the effects of sowing density of wheat on **a)** grain yield and **b)** the mass of weeds per square metre.

Intraspecific competition

We will look at some more data involving **sowing density**. Table 7 shows the results of a trial to find the effect of sowing density of cotton on the mean height of the plants.

Mean number of seeds sown per hectare	Mean height of cotton plants/m	Standard deviation	Standard error
67 000	1.15	0.18	0.06
33 000	1.27	0.11	0.04
17 000	1.31	0.11	0.04

Table 7 Effect of sowing density on the mean height of cotton plants.

1 The cotton seeds in the first row in the table were sown at a density of 67 000 seeds per hectare. What was the mean number of seeds sown per square metre?

2 What do the figures in the first two columns of the table suggest about the effect of sowing density on the height of cotton plants? Suggest an explanation for your answer.

Plot the information in this table as a simple graph similar to that in Figure 12. Plot the mean values for each sowing density as a cross and draw bars to represent two standard errors on either side of these mean values.

3 Use your graph to explain whether sowing density has a significant effect on the height of cotton plants.

4 Look at the formula for standard error again.

$$SE = \frac{SD}{\sqrt{n}}$$

Explain what will happen to the value of the standard error if you were to increase sample size.

5 The scientists who carried out this investigation planned to repeat it the following year on a larger scale. Use your answer to Q4 to suggest why.

Squirrels and interspecific competition

There are two species of squirrel in the UK – the native red squirrel and the introduced grey squirrel. Red squirrels are smaller than grey squirrels. They store less body fat and they spend more time in the tree canopy. Scientists have suggested several hypotheses to explain why grey squirrels have replaced red squirrels in much of southern Britain. One hypothesis is that grey squirrels are better adapted to living in oak woods. Acorns are the fruit of oak trees. They mature in the autumn and fall to the ground where squirrels can collect them. Earlier research showed that grey squirrels are much better than red squirrels at digesting acorns.

Look at the graph in Figure 16. It shows the relationships between the numbers of the two species of squirrel and acorn numbers.

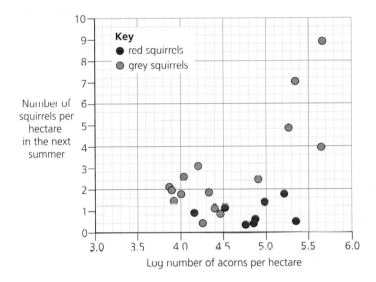

▲ **Figure 16** Graph showing the relationship between number of acorns produced per hectare and number of squirrels the following summer.

■ **Why are the figures for the number of acorns given on a log scale?**

A log scale increases in multiples of 10. On the scale shown here a log value of 3 represents 1000; a value of 4 represents 10 000; a value of 5 represents 100 000 and a value of 6 represents 1 000 000. Using a log scale lets us plot a much larger range of values on a single axis.

■ **Suggest how the data relating to the number of acorns might have been collected.**

The technique used would have to be based on a sampling method. Quadrats were probably used. This would allow the number of acorns in a given area to be found. This figure could then be converted to the number of acorns per hectare.

■ **What is the relationship between the log of the number of acorns per hectare and the number of grey squirrels the next summer?**

If you were to plot a curve of best fit, you would get a line sloping upwards. This shows that there is a positive correlation – the greater the log of the number of acorns per hectare, the greater the number of grey squirrels the next summer.

■ **What is the relationship between the log of the number of acorns per hectare and the number of red squirrels the next summer?**

Plotting the curve of best fit this time gives a line parallel to the *x*-axis. This shows that there is no correlation.

■ **Use the information in this section to suggest how interspecific competition could account for the absence of red squirrels in most of southern England.**

Much of the native woodland in southern England is oak. Grey squirrels are more successful in oak woods because they spend more time on the ground where acorns fall. They can also digest the acorns better and put on more body fat as a result. A greater amount of body fat probably enables grey squirrels to survive over winter in oak woods better than red squirrels.

Predators and their prey

▲ **Figure 17** A spider with a grasshopper in its web – an example of a predator and its prey.

Predators are animals that kill and eat their **prey**. Figure 17 shows a spider that is feeding on a grasshopper. The spider is the predator and the grasshopper is its prey. Obviously, the number of predators influences the number of prey – as the number of predators increases, the number of prey decreases. Equally obviously, the number of prey influences the number of predators. Without enough food, fewer predators will survive.

Scientists use models to help them understand the relationship between the population of a predator and the population of its prey. We will look at one of these models, called the Lotka-Volterra model after the two scientists who originated it (Figure 18).

- Look at the part of the graph in Figure 18 between points **A** and **B**. It shows the prey population increasing. Therefore there will be more food available for the predator population. The predators will breed successfully and their numbers will increase.
- With more predators, however, more prey will be killed so we see, between points **B** and **C**, a fall in prey numbers. There is now less food for the predators and their population starts to fall.
- Numbers of prey rise again and the pattern will be repeated.

We will look at a piece of evidence that appears to support this model. Figure 19 shows a graph of changes in the numbers of skins of a predator, the lynx, and the number of skins of its prey, the snowshoe hare, bought from trappers by the Hudson Bay Company, Canada for the years 1890 to 1920.

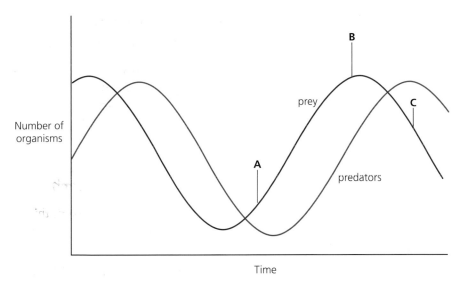

▲ **Figure 18** The Lotka-Volterra model predicts that the populations of a predator and its prey change over time.

■ Describe how the curve for the number of lynx skins is different from the curve for the number of snowshoe hare skins for the period shown in Figure 19.

There are two features to note here. The peak for the lynx skins always comes a little later than the peak for the snowshoe hare skins and the lynx numbers are always lower than the hare numbers.

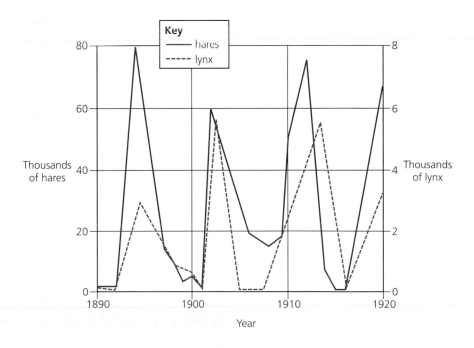

▲ **Figure 19** Graph showing the numbers of lynx and snowshoe hare skins bought by the Hudson Bay Company between 1890 and 1920.

■ **Explain the change in the number of snowshoe hare skins purchased between 1908 and 1912.**

Lynx numbers are low, as shown by the number of skins bought by the Company. With few predators to kill them, snowshoe hare numbers rise.

■ **In one population cycle, lynx numbers rise after snowshoe hare numbers have risen. Explain why.**

The rise in lynx numbers is the result of more prey being available. Therefore their numbers will only rise after snowshoe hare numbers have increased.

■ **Explain what happened to the lynx numbers between 1910 and 1917.**

The rise in the lynx numbers caused a fall in the numbers of snowshoe hares. There is, therefore, less food for the lynx and, as a result, their numbers fall.

The changes in the numbers of snowshoe hare skins and lynx skins shown in Figure 19 seem to support the Lotka-Volterra model but more recent experimental work has shown that the situation is really much more complicated than this simple model suggests. Here is some information resulting from recent ecological investigations involving snowshoe hares and lynx.

- High populations of snowshoe hares are associated with a shortage of plant food on which the animals feed. When the snowshoe hare population is at its lowest, the plants start to grow again but, for the next two or three years, toxins in the young shoots are thought to delay the next rise in the population of snowshoe hares.
- When snowshoe hare numbers decrease, lynx eat other prey animals. There is very little evidence of lynx dying of starvation.
- Scientists attached radio collars to lynx. They showed that when the number of snowshoe hares was low, the lynx tended to move away – in one case, they travelled as far as 800 km.
- There are no lynx on Anticosti Island in Eastern Canada. The population of snowshoe hares on the island, however, still shows a regular population cycle.

What we can see from this information is that simple models, such as the Lotka-Volterra model, provide a very useful starting point for analysing population cycles. They cannot provide us with a complete explanation.

Human populations

Modern humans are thought to have become distinct from their ancestors about a million years ago. At first they were nomadic, and lived as hunters of animals and gatherers of wild plant food. In this early period of history, human populations increased slowly to reach an estimated 5 to 10 million. Following the development of agriculture and the domestication of animals,

Q7

Why might it be better to use log scales rather than arithmetic scales to plot data on human population growth?

Q8

Birth rate is always measured per year. It would be more difficult to compare birth rates if they were calculated for shorter periods of time. Explain why.

human populations grew more rapidly but had still only reached about 800 million people by 1750. In the short period of time from 1750 to the present day, the total number of humans has risen to approximately 6.7 billion. More than 80% of the increase in human numbers has taken place in less than 0.1% of human history!

Calculating the rate of human population growth

If we ignore the effect of immigration and emigration, then the difference between birth rate and death rate determines the rate of growth of a population. If birth rate is greater than death rate, the population increases. If death rate is greater than birth rate, the population decreases.

There are various ways of measuring birth rate, but the simplest measure is **crude birth rate**. This is given by the formula:

$$\text{crude birth rate} = \frac{\text{number of births in a particular year}}{\text{total population in that year}} \times 1000$$

Multiplying by 1000 gives the rate per thousand of the population.

We calculate crude death rate in a similar way.

$$\text{crude death rate} = \frac{\text{number of deaths in a particular year}}{\text{total population in that year}} \times 1000$$

Calculating population growth rate is a simple matter of subtracting crude death rate from crude birth rate. This will give population growth rate per 1000 of the population. We will look at an example. Table 8 shows some data for West Africa in 1992.

Total population at the start of 1992	182 000 000
Crude birth rate for 1992	47 per 1000
Crude death rate for 1992	17 per 1000

Table 8 Data relating to population, birth rate and death rate for West Africa in 1992.

■ Use the data in the table to calculate the population growth rate.

We subtract crude death rate from crude birth rate, so:

$$\text{population growth rate} = \text{crude death rate} - \text{crude birth rate}$$
$$= 47 - 17$$
$$= 30 \text{ per } 1000$$

■ By how much did the population of West Africa increase in 1992?

We have just calculated the population growth rate as 30 per 1000 and we know that the total population at the start of 1992 was 182 000 000. Therefore the total population grew by:

$$30 \times \frac{182\,000\,000}{1000} = 5\,460\,000$$

■ Calculate the percentage growth rate for 1992.

$$\text{percentage growth rate} = \frac{\text{population increase during 1992}}{\text{population at start of 1992}} \times 100\%$$

We found the population increase in the previous calculation, so:

$$\text{percentage growth rate} = \frac{5\,460\,000}{182\,000\,000} \times 100\%$$

$$= 3\%$$

■ How reliable are the results of these calculations?

The results of calculations such as those above are only as reliable as the data on which they are based. In many of the world's poorest countries, such as some of those in West Africa, data are only available for part of the population.

Data such as those in Table 8 are usually collected on a national or a regional basis. Therefore we need to bear in mind that, although the crude death rate in West Africa as a whole, in 1992, was 17 per 1000, it varied widely from country to country. It was only 8 per 1000 in the Cape Verde Islands, but 23 per thousand in Sierra Leone. Even within a country or a small community, there are variations both in crude birth rate and in crude death rate.

Crude death rate

We will consider variation in death rate in a small village in West Devon. In this village, it was thought that there was a significant difference in death rate due to cancer north and south of a river running through the village. Scientists collected the data shown in Table 9.

	North of the river	South of the river
Number of deaths from cancer in 100 deaths from all causes	26	12

Table 9 The results of an investigation into the number of people dying from cancer on either side of a river flowing through a West Devon village.

We use the χ^2 (chi-squared) test when the measurements we have collected involve the number of individuals in particular categories. The χ^2 test is based on calculating the value of χ^2 from the equation:

$$\chi^2 = \sum \frac{(O - E)^2}{E}$$

Where O represents the observed results and E represents the expected results.

- Start by producing a null hypothesis. With the χ^2 test the null hypothesis is always 'There is no difference between the observed results and the expected results'.

- Work out the values of the observed results and the expected results. The results in the table give us the observed results (O). We have to calculate the expected results (E). If dying of cancer was just a matter of chance, we would expect the number of people dying of cancer north of the river to be the same as the number of people dying of cancer south of the river. So the expected numbers dying of cancer are:

$$\frac{26 + 12}{2} = 19$$

- Calculate χ^2. We now use the equation to calculate the value of χ^2. The best approach is to write the values in a table (Table 10).

	Deaths from cancer in people living	
	north of river	south of river
Observed results (O)	26	12
Expected results (E)	19	19
$(O - E)^2$	49	49
$\dfrac{(O - E)^2}{E}$	2.6	2.6
$\sum \dfrac{(O - E)^2}{E}$	5.2	

Table 10 Calculating χ^2.

In the equation, the symbol 'Σ' means 'the sum of' so in the last row of the table we added together all the separate values of $(O - E)^2/E$.

- Look up and interpret the value of χ^2. The confidence you can place in your judgement will depend on how many categories your data fall into. **Degrees of freedom** take this into account. The number of degrees of freedom is one fewer than the number of values of $(O - E)^2/E$ that are calculated. In this case, it is 1.

We pointed out on page 9 that, for biological investigations, if an event fails to happen only once in twenty times, in other words, it has a probability of 0.05 or less ($p = 0.05$), we will accept this as support for our null hypothesis. We call $p = 0.05$ the **critical value**. We now look up the value for χ^2, for 1 degree of freedom, in Table 11.

With 1 degree of freedom, the value of χ^2 that we calculated, 5.2, is greater than the critical value of 3.84. We can therefore reject the null hypothesis and say that there is a significant difference between the observed and expected values.

Degrees of freedom	Critical value
1	3.84
2	5.99
3	7.82
4	9.49
5	11.07
6	12.59
7	14.07
8	15.51
9	16.92
10	18.31

Table 11 Table showing the critical values of χ^2 for different degrees of freedom.

The demographic transition

As a country develops, its total human population increases, and its birth rate and death rate change. One of the models used to describe this pattern of change is called the **demographic transition**. It shows how economic development from a mainly rural, largely illiterate society that is dependent

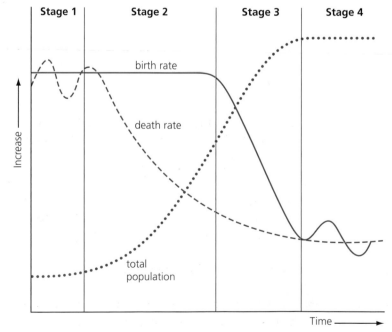

Stage 1 High stationary stage:
Stable population; high birth rate; high death rate. There is a high birth rate but because the food supply is limited and infectious disease is common, there is also a high death rate. Overall, there is little population growth.

Stage 2 Early expanding stage:
Increasing population; high birth rate; falling death rate. Improvements in agriculture and sanitation lead to a fall in death rate. The birth rate remains high so the population increases.

Stage 3 Late expanding stage:
Increasing population; falling birth rate; low death rate. The death rate is low and fairly stable. The birth rate starts to fall. The population continues to increase, but at a slower rate.

Stage 4 Late stationary stage:
Stable population; low birth rate; low death rate. The death rate is low and fairly constant. The birth rate is also low but fluctuates with changing economic conditions. The population shows little change.

▲ **Figure 20** The main features of the demographic transition.

Unit 4 Populations and environment

on farming, to an urban, literate and industrial one, is accompanied by a series of changes. During the course of these changes, death rate falls leading to a surge in population growth. The fall in death rate is followed by a fall in birth rate. The population changes from a low stable population with high birth and high death rates to a high stable population with low birth and death rates. The main features of the demographic transition are summarised in Figure 20.

Graveyards and life tables

We can get some idea of the characteristics of the UK population during the early stages of this model by looking at some of the evidence from graveyards. Inscriptions on gravestones tell us a lot about individual humans but they also provide information about human populations. The photograph in Figure 21 shows part of a large graveyard in Northumberland and Table 12 contains data about the number of deaths between the years 1750 and 1800 collected from this graveyard.

▲ **Figure 21** Part of a large graveyard in Northumberland. Data collected from individual gravestones are summarised in Table 12.

Age interval in years	Number of people dying in age interval
0–9	24
10–19	6
20–29	4
30–39	5
40–49	4
50–59	2
60–69	5
70–79	3
80–89	3

Table 12 Data collected from gravestones referring to age at death for the period 1750–1800.

We can use the data in Table 12 to construct a life table (Table 13). Before we look at any of the data in the table, we will look carefully at the column headings.

Column 1 The oldest person buried in the period 1750–1800 lived to be 88 years old. We need to group the ages or we will end up with too many rows in our life table to analyse and it would be very difficult to see patterns. In this life table, we have grouped the ages in ten-year intervals.

Column 2 We have used the data from the second column of Table 12 to complete this column.

Column 3 We calculate the figures for this column by adding the figures in column 2 starting at the bottom. We will look at the 60–69 age group as an example. The number of people alive at 70 years old were the three who died between 70 and 79 and the three people who died between 80 and 89, making a total of six. Similarly the number of people alive at 60 could be calculated from $5 + 3 + 3 = 11$.

Column 4 This column is an intermediate step that we need in order to calculate the figures in column 5. To calculate the figures in column 4 we add those in column 3 from the bottom up.

Column 5 This is the average number of years lived from the mid-point of the age interval. The figures in this column are calculated from the equation:

$$e_x = \left(\sum l_x \times 10\right) - 5$$

In this equation, we multiply by 10 to convert age intervals of 10 years into years. If we expect a person to live a further 1.4 age intervals, this would be 14 years. We take off 5 because this is half the age interval and we would expect, on average, a person to live halfway through their final 10-year age interval.

Column 6 The figures in this column are calculated by dividing the number of people alive in the next age interval (l_{x+1}) by the number alive in this age interval (l_x).

1 Age interval in years	2 Number of people dying in age interval	3 Number of people alive at start of age interval	4 Number of people alive in next age interval if none had died after reaching this age interval	5 Life expectancy in years	6 Probability of surviving from one age interval to the next
x	d_x	l_x	$\sum l_x$	e_x	p_x
0–9	24	56	186	28.2	0.57
10–19	6	32	130	35.6	0.81
20–29	4	26	98	32.7	0.81
30–39	5	22	72	27.7	0.77
40–49	4	17	50	24.4	0.76
50–59	2	13	33	20.4	0.85
60–69	5	11	20	13.2	0.55
70–79	3	6	9	10.0	0.50
80–89	3	3	3	5.0	0.00

Table 13 A life table constructed from data shown in Table 12.

We will look at some of the features of this life table.

■ What do the data in the life table tell us about the children in the age group 0–9 years?

Column 2 shows that there are more deaths in this age group than in any other. This is supported by the data in column 6. They show that the probability of surviving into the next age group is relatively low ($p_x = 0.57$).

■ At what age would you expect a 5-year-old child to die?

The mid-point of the 0–9 age group is 5 years. Look at column 5 and you will see that the average number of years lived from this point is 28.2. Add 5 to that and we get a figure of 33.2 years.

■ Suggest why the figure for column 6 is low between 0 and 9 years old.

Birth and early post-natal life were dangerous times. Infections killed many children. Malnutrition was also significant.

■ Suggest why the figures for column 6 are low after 80 years old.

Older people are vulnerable to infections and other conditions associated with old age. None of the people in this small sample lived beyond 89.

The quality of data

Values such as those calculated in Table 13 can look very impressive but it is important to remember that statistical analyses are only as reliable as the data on which they are based. Were these data a representative sample?

- The sample was very small, only 56 people altogether.
- In the period of time involved, only richer people could afford to erect gravestones. Many poor people were buried without gravestones.
- Most of the people buried in this graveyard lived locally in the town. Nobody living in the surrounding countryside was buried here.

We can use data from life tables to plot survival curves and construct population age pyramids.

Survival curves

When you look in an old graveyard, such as that shown in Figure 21 (page 25), one of the most striking features is the number of young children buried there. The data in Table 12 show that 24 out of the 56 people whose deaths were recorded died before they were 10 years old. Suppose we had started with a population of 1000 instead of 56. We would expect:

$$24 \times \frac{1000}{56} = 429$$

to have died before they reached their tenth birthday. In other words, only 571 would have survived. Similar calculations show the number of people

Q9

Suppose poor people had been included in this sample. How would you have expected the data in column 6 of Table 13 to have been different? Explain your answer.

1 Populations

Q10

What is the average life expectancy for the population shown in Figure 22?

Q11

A survival curve for people living in the same area today would have a very different shape from that shown in Figure 22. Describe how the shape would be different.

we would expect to be still alive after 20 years, 30 years and so on. We can plot the results of these calculations as the **survival curve** shown in Figure 22. The **average life expectancy** (the age at which 50% of the population are still alive) may be found from this curve.

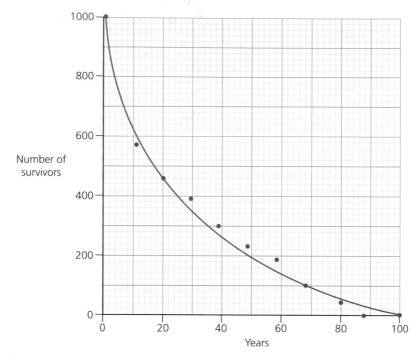

▲ **Figure 22** A survival curve for people living in a region of Northumberland between 1750 and 1800. A curve of best-fit has been drawn.

The change in the shapes of the survival curves for human populations at different stages in the demographic transition may be linked to changes in the cause of death (Figure 23).

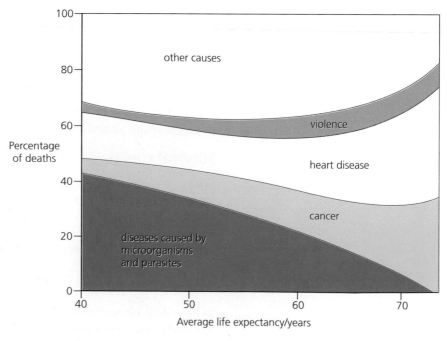

▲ **Figure 23** Likely causes of death for populations with different average life expectancies.

Populations such as that represented in Figure 22 and those of developing countries are at an early stage in demographic transition and have a low average life expectancy. Populations from developed countries are at a later stage in demographic transition – they have a high average life expectancy. The data in Figure 23 show that infectious disease causes over 40% of deaths at an early stage in demographic transition. Many of these diseases affect children and this is one of the main causes of a low average life expectancy. Cancer and disease of the circulatory system are the main causes of death at later stages in the demographic transition. They mainly affect older adults.

When interpreting these data and applying the information in the diagram, there are some important points that you should consider.

- The data for cause of death are given as percentages. Think about what this means for deaths from, for example, cancer in a population whose average life expectancy is 70 years. You would expect an increase in the number of cases of cancer with increased life expectancy because people who live longer are more likely to develop cancer. If there are more deaths from cancer, it is likely that the percentage of people dying from the condition will also be high. Fewer people, however, die from infectious disease. This could also explain the large percentage of people dying from cancer, since they must die of something, and they haven't been killed by infectious disease. An increase in the percentage of people dying of cancer could then be due to more people developing cancer or to fewer people dying of infectious disease.
- The data on which Figure 23 is based refer to 1962. Death from other causes features prominently. One of these other causes was diarrhoea in infants. Since 1962 there have been many advances in medical knowledge. One of these advances was the discovery that dehydration from infantile diarrhoea could be treated cheaply and effectively with oral rehydration solutions. Fewer young children now die from infantile diarrhoea. We should bear in mind that the data that you find in books and on the Internet are not always up to date and this may invalidate some of the conclusions that can be drawn from them.

Population age pyramids

A histogram is a convenient way of plotting suitable data about the age of people in a population (Figure 24, overleaf).

If we display a histogram for males horizontally and add to it a mirror image of a similar histogram for females, we have an **age population pyramid** or, more simply, an **age pyramid**. Age pyramids show the percentage of males and females in a population at any one time (Figure 25).

In a developed country such as the UK:

- A small percentage of the population consists of children under 15.
- The sides of the pyramid are more or less vertical. There are some small bulges and narrower parts. These usually represent a lower number of births in times of economic hardship.
- A large percentage of the population is over 65.

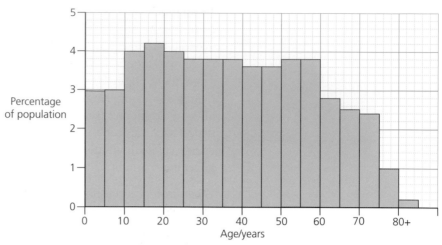

▲ **Figure 24** The number of males in different age groups in the UK population in 1995.

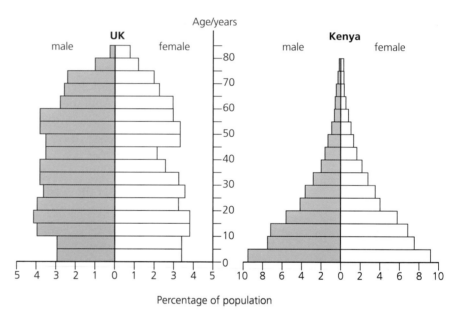

▲ **Figure 25** The main features of age pyramids for developed countries (UK) and developing countries (Kenya).

Q12

The population of some countries is decreasing. Describe the main features of an age pyramid representing a decreasing population.

In a developing country such as Kenya:

- A large percentage of the population consists of children under 15.
- The sides of the pyramid slope in steeply.
- A small percentage of the population is over 65.

Succession

Some of the species of organisms living in a community are gradually replaced by others. This is called **succession**. It is an ecological process resulting from the activities of the organisms themselves. Over a period of

▲ **Figure 26 a), b)** Succession is an important ecological process associated with slow-flowing streams and rivers. Some of the plants in the foreground of photograph **a)** have leaves that float on the surface. Their roots trap particles of silt carried in the water and this forms mud. As the water becomes shallower, these floating plants are replaced by upward growing reeds and other plants.

The first organisms to colonise the bare rock in photograph **b)** are lichens but they are only able to grow when water is available. As the soil that begins to accumulate round them is often washed away, succession is very slow. Ultimately though, the lichens will be replaced by mosses. These, in turn, will give way to ferns and species of flowering plant.

time, the organisms modify their environment. These modifications produce conditions better suited to the growth of other species (Figure 26).

Succession on dead bodies

A dead body gradually decomposes. As it does, it attracts a variety of insects. These insects feed and lay their eggs on the decomposing tissues (Figure 27). A forensic entomologist studies the insects present on a decomposing body and can estimate the length of time the person has been dead. Bodies of all terrestrial mammals decompose in the same way. The main stages in decomposition are shown in Table 14 on the next page.

▲ **Figure 27** Blowfly maggots feeding on a carcass. On a warm summer's day female blowflies may arrive within minutes of the animal's death and lay their eggs within half an hour.

Stages in decomposition

Time after death/days	Stage	Notes
0–2	Fresh	The body appears fresh but, inside, bacteria have already begun the process of decomposition.
2–4	Bloated	The body swells because of gases produced by bacteria in the digestive system. There is a smell of decomposition.
4–8	Active decay	Exposed parts of the body turn black as a result of aerobic breakdown of protein. The body smells strongly of ammonia.
9–12	Advanced decay	The body begins to dry out. There is still some soft tissue. This contains fats that have not yet decomposed. There is a smell of butyric acid.
Over 13	Dry remains	The body is almost dry. Various amounts of hair and skin still cover the bones.

Table 14 The main stages in the decomposition of a dead mammal.

Forensic entomologists have found that bodies at different stages of decomposition attract different species of insects.

1 a) Some species of beetle are attracted to a dead body at the active decay stage. Use the information in the table to suggest what attracts these beetles at this stage.

b) Describe briefly an experiment that would allow you to investigate whether your suggestion was correct.

2 It is important that data on insect succession collected in one geographical area are not used for finding the time of death in another geographical area. Suggest why.

Environmental factors influence insect succession. These factors include temperature.

3 Explain how temperature may affect insect succession in a dead body.

A research worker investigated the populations of two species of beetle, **A** and **B**, on deer carcasses in Florida. She counted the number of beetle larvae on a total of seven carcasses. Her results are shown in Figure 28.

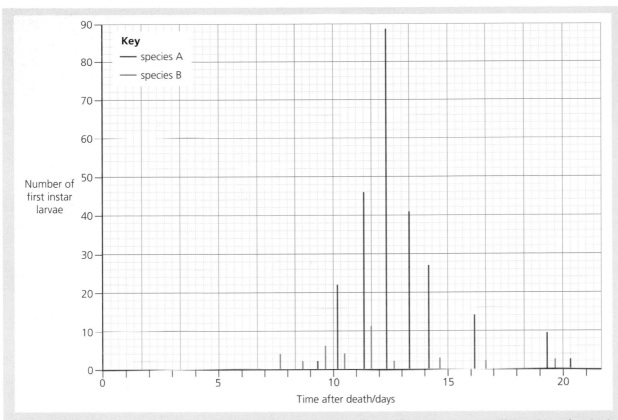

▲ **Figure 28** Changes in the numbers of first instar larvae of two species of beetle on deer carcasses at different times after death.

The body of an insect larva is surrounded by a tough exoskeleton. When the larva grows, it moults its old exoskeleton. It increases in size while the new exoskeleton, formed underneath the old one, is still soft. The stage between successive moults is called an instar.

4 All the larvae in Figure 28 were in their first instar. Explain why it was important that they were in the same instar.

5 A wildlife officer in Florida found a dead deer. Could he use the information shown in Figure 28 to produce a reliable estimate of its time of death? Explain your answer.

Succession in sand dunes

Sand dunes occur in many coastal areas and are a good example of succession. Succession is a gradual process and the time scale involved prevents us from sitting in one place and observing successional changes. If we walk, however, from the sea shore in to the sand dunes we will pass through different areas which represent different stages in succession (Figure 29, overleaf).

We will start nearest the sea on a sandy shore, above high tide level. Some plants, such as sea couch grass and lyme grass, are able to colonise and survive in the bare sand. Their roots form a dense network that binds sand particles together and, as a result, sand starts to pile up and form the **fore dunes**. As the plants grow larger, their aerial parts trap more and more sand. They cannot grow fast enough, however, to avoid being smothered by sand when it is piling up at a rate faster than $30\,\text{cm year}^{-1}$. Under these conditions they are replaced by marram grass.

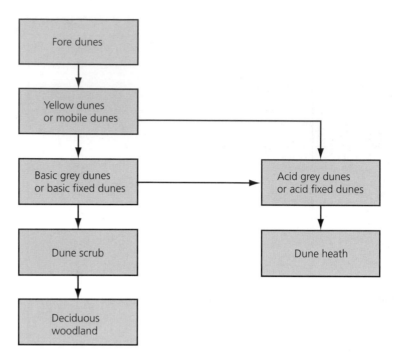

▲ **Figure 29** Succession in sand dunes.

Marram-grass dominated dunes are called **yellow dunes** or **mobile dunes** – yellow because there is very little humus in the sand and it shows yellow through the marram grass cover; mobile because they are continually changing shape as the wind scours the face and blows the sand. This is a very harsh environment and few plants can survive in these conditions. Look at Table 15. This table compares the abiotic conditions in these yellow dunes with the conditions in the grey or fixed dunes later in the succession.

Abiotic factor	Yellow dunes	Grey dunes
Mean wind velocity 5 cm above the dune surface/km h^{-1}	12.1	2.4
Organic matter/%	0.3	1.0
Concentration of sodium ions/ppm	8.5	4.2
Concentration of calcium ions/ppm	637.0	297.0
Concentration of nitrate ions/ppm	48.0	380.0

Table 15 A comparison of some of the abiotic factors in yellow and grey dunes.

The high wind speed has two important effects. It will pile sand on top of any plants growing there and it will lead to a high rate of water loss by transpiration. There is very little organic matter, so the sand does not retain water very well. It will dry out rapidly after rain. Important soil nutrients such as nitrogen are in short supply, but sea spray results in high concentrations of sodium and calcium ions. Very few plants can grow here but one that does is marram grass. We call species such as sea couch grass, lyme grass and marram grass **pioneer species** because they are the first plants to colonise the area. Marram grass has a number of adaptations that enable it to grow in these conditions (Figure 30).

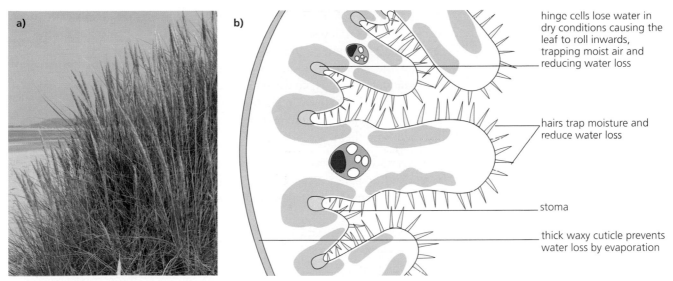

hinge cells lose water in dry conditions causing the leaf to roll inwards, trapping moist air and reducing water loss

hairs trap moisture and reduce water loss

stoma

thick waxy cuticle prevents water loss by evaporation

▲ **Figure 30 a)** On sand dunes, marram grass is a pioneer species. Its leaves grow from a vertical underground stem, so remain above the sand that the wind deposits on the plants. It also has adaptations that enable it to grow in areas where there is often little water available. Some of these adaptations can be seen in the section through a leaf shown in **b)**.

Marram grass illustrates an important biological principle:

The harsher the environment, the fewer the species that are able to survive and the lower the species diversity. In harsh environments, it is generally abiotic factors that determine the species present.

Now we will go inland to the area of **grey dunes** or **fixed dunes**. They are called grey dunes because humus in the sand colours them grey; they are referred to as fixed dunes because the sand is no longer being blown about, so they are much more stable. Table 15 shows that abiotic factors have changed and many of these changes are due to the activity of organisms on the yellow dunes. The roots of marram grass bind the particles of sand and the leaves act as a windbreak. The wind velocity is lower so less sand is blowing about. Dead material falls from the marram grass and is broken down by soil bacteria. The amount of humus is higher, so the developing soil retains moisture better and the concentration of important soil nutrients such as nitrates rises. Rain is also beginning to leach the soluble sodium and calcium ions from the surface layers of the soil. Other plants can now grow and they gradually replace marram grass. We are beginning to arrive at a situation where:

The environment is less harsh. There are more species and a greater species diversity. In this environment it is biotic factors, such as competition, that determine whether particular species survive.

The process of succession continues. The species found on these grey dunes change their environment in such a way that they are replaced by other species. **Dune scrub** starts to develop. Hawthorn and elder grow and shade out the shorter vegetation. Ultimately woodland develops. We have reached a stage where no further change takes place. This is the **climax community**. In Britain, it is usually woodland of some sort.

If you look again at Figure 29 (page 34), you will see that not all succession on sand dunes follows this pattern. The sand particles that make up the dunes have different origins. Where they originate from rocks such as granite, the soils they produce are more acidic. They support different plants and gradually give way to dune heath dominated by plants such as bracken and heather. Basic soils containing high concentrations of calcium carbonate can also become more acidic as soluble basic ions are gradually leached from them.

The details of the processes of succession in slow-flowing streams and rivers and on bare rock shown in Figure 26 are different from those on sand dunes but the principles are very similar. They are summarised in Figure 31.

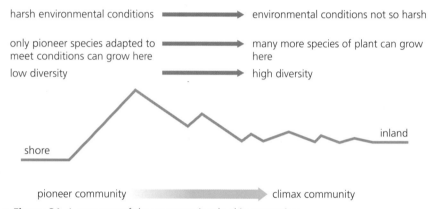

▲ **Figure 31** A summary of the processes involved in succession.

Succession and human activity

Succession does not always proceed all the way to a climax community. It can be stopped by various factors often associated with human activity. In an investigation, ecologists compared control plots with plots on which the plant cover had been partly destroyed by allowing motor cycles to be ridden over it. The results are shown in Figure 32.

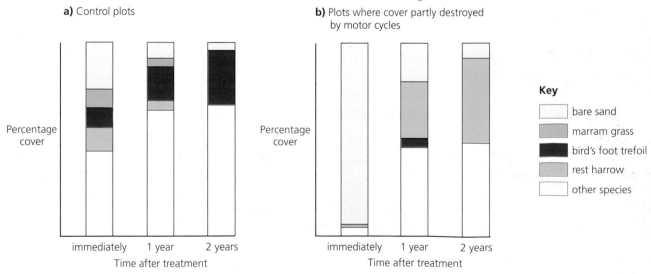

▲ **Figure 32** A comparison of changes in the vegetation of **a)** control plots and **b)** plots in which the plant cover had been partly destroyed by motor cycles.

■ **Describe how the control plots should have been treated.**

In this case, the important thing was to treat the control plots in the same way as the experimental plots in all aspects except in riding a motor cycle over them.

■ **What was the purpose of the control plots?**

They provide a comparison with the experimental plots by showing what would happen to similar areas of dune that were not damaged by being ridden over.

■ **Describe the process of succession that takes place in the control plots over the period covered by this investigation.**

There is a decrease in the amount of bare ground and a change in plant cover. The percentage cover of marram grass and rest harrow decreases and there is an increase in bird's foot trefoil and other species.

This investigation illustrates an important point. When we halt the process of succession, in this case by partly destroying plant cover with a motor cycle, it rarely returns to where it began.

■ **Use the results in the bar chart to describe how the process of succession is different after the plant cover has been destroyed. Suggest an explanation for the difference.**

In a natural succession, marram grass is a pioneer species. In the investigation, we are starting from a position where species other than marram grass are present, although most of the plant cover has been destroyed. The sand in the areas from which the plant cover has been removed is quite different from the sand in which the marram grass became established initially. The sand with reduced plant cover contains more humus and it is not so likely to dry out. It also has a lower concentration of sodium and calcium ions and a higher concentration of nitrate ions. These are conditions in which rest harrow grows better than marram grass. You can see quite clearly how human activity has altered the path of succession.

Managing chalk grassland

There is little, if any, of the landscape of southern Britain that has not been modified by human activity. Much of our current landscape is the result of centuries of agricultural practice. Chalk grassland is one example. It consists of a diverse mixture of grasses and herbs and can support up to 50 species of flowering plant per square metre (Figure 33). There is also a high diversity of invertebrates, such as insects and snails.

At the beginning of the nineteenth century more than 50% of the South Downs was chalk grassland. Today, the figure stands at just 3%. There are many reasons for this massive reduction. Understanding them requires not

▲ **Figure 33** Orchid-rich chalk grassland is identified as a European conservation priority.

only a knowledge of ecology, but also an understanding of other aspects of biology.

- Chalk grassland resulted from sheep grazing the thin, nutrient-poor soils overlying chalk. A massive decline in the number of sheep and the resulting spread of scrub has reduced the amount of chalk grassland. The bushes that form this scrub also spread to remaining areas of chalk grassland. These bushes, however, are also part of the chalk ecosystem and are important in contributing to the overall biodiversity by providing a habitat for many species. A balance needs to be achieved by controlling this process of succession.
- The disease myxomatosis wiped out large numbers of rabbits in the 1950s. Since then rabbits have become resistant to the disease and numbers have increased dramatically. They have little effect in controlling scrub vegetation and their warrens and overgrazing cause serious erosion.
- In order to improve yield, many farmers have added nitrogen-containing fertilisers and selective herbicides to fields. Some species have benefited and there has been a resulting loss in biodiversity.
- Much of the existing chalk grassland is fragmented. This fragmentation causes isolation and makes populations more vulnerable to local extinction from disease and predation.

*Q*13

Use your knowledge of competition to explain why the use of nitrogen-containing fertilisers on chalk grassland has resulted in a loss of biodiversity.

Chapter 2
Energy transfer

The concentration of carbon dioxide in the Earth's atmosphere is increasing. Scientists have collected evidence which suggests that this increase is leading to global warming and climatic change. A relatively small increase in mean global temperature could have major effects on communities of living organisms. Computer models can help biologists to predict these effects. We will start this chapter by looking at an early model that was used to predict the effect of an increase in carbon dioxide concentration on forests in North America.

Figure 1 shows a map of the main vegetation zones in North America. In the far north there is tundra which is low-growing vegetation adapted to the short Arctic summer and able to withstand the long cold winters. Further south there is a belt of coniferous forest. Here the summers are cool and wet, but the winters are still very cold and last for at least six months. Further south still summers become warmer and winters much shorter. Here, the vegetation depends to a large extent on rainfall. In the drier west there are grass-covered prairies. In the wetter east there are deciduous forests.

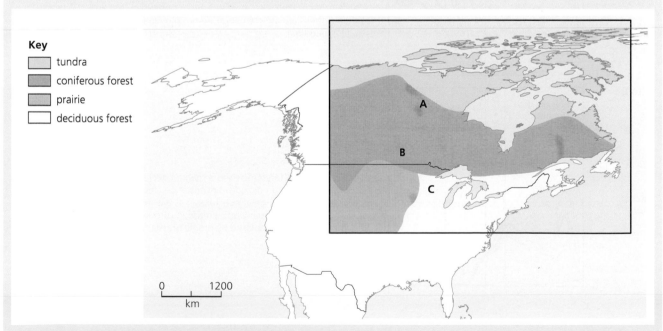

Key
- tundra
- coniferous forest
- prairie
- deciduous forest

0   1200
km

▲ **Figure 1** A map showing the main vegetation zones in the northern part of North America. Scientists ran computer simulations for sites **A**, **B** and **C** to predict the effects of increasing carbon dioxide concentration and climatic change.

Before they ran their computer simulation, the scientists input data about likely trends in carbon dioxide concentrations and climate. They assumed that:

- forest at all three sites started growing on bare soil
- carbon dioxide concentration in the atmosphere was 0.03%
- climatic conditions were those in the mid-1980s when the work was carried out.

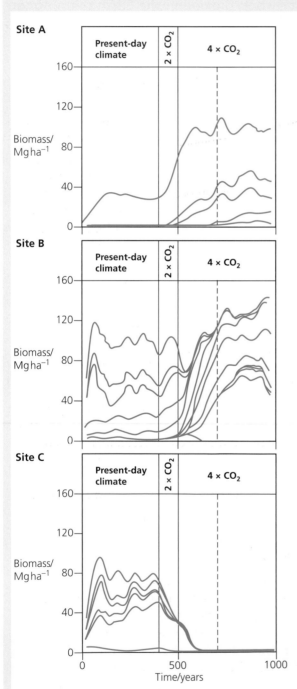

Site **A** (northern edge of coniferous forest) The *y*-axis shows biomass in $Mg ha^{-1}$. Over the first 400 years the biomass increases as the forest becomes established and then it becomes more or less constant. The graph also shows that almost all the trees present are of one species. This is spruce. The changes in carbon dioxide concentration result in a greatly increased biomass and a greater number of species. Not only are there conifers, such as spruce, but deciduous trees, such as birch.

Site **B** (coniferous forest) Here the effects of increasing carbon dioxide concentration are more complex. The total biomass doesn't change very much but the species composition changes quite a lot and there is a much greater diversity of trees as a result of increasing carbon dioxide concentration. Look also at what happens between 500 and 700 years. Different trees are affected at different times so the overall changes take quite a long time.

Site **C** (deciduous forest) There is a marked decrease in biomass that starts as soon as the carbon dioxide concentration starts to increase. This is because climatic change affects rainfall as well as temperature. The drier conditions that accompany an increase in carbon dioxide concentration result in forest being replaced by prairie grassland.

▲ **Figure 2** The results of a computer simulation showing likely changes in forests at points **A**, **B** and **C** as a result of the changes in carbon dioxide concentration shown.

From this starting point they suggested the following sequence of events.

- The carbon dioxide concentration remained at 0.03% for the next 400 years.
- Between 400 and 500 years the carbon dioxide concentration gradually doubled and climatic conditions were adjusted to account for this.
- Between 500 and 700 years, carbon dioxide concentration doubled again to a level four times the original. Again, climatic factors were adjusted in line with this.
- After 700 years, the carbon dioxide concentration remained constant at a level of four times the original for another 300 years.

The data for carbon dioxide concentration are shown in Figure 2, together with the results of the computer simulation.

Computer models are much more reliable if they can be validated. At the end of the last glaciation, warmer conditions developed. These conditions also resulted in changes to North American forests. Other scientists have followed the changes that took place at the end of the last glaciation by examining samples of pollen preserved in peat deposits. Their results were very similar to the predictions of the computer model we have studied above. This suggests that the computer model is valid. Models such as this one help scientists to understand the ecological consequences of climatic change.

In this chapter we look more closely at the issues surrounding energy transfer and climate change. Plants and other chlorophyll-containing organisms photosynthesise, using carbon dioxide and water, to produce carbohydrates. In this process energy from sunlight is transferred to chemical potential energy in glucose and other carbohydrates.

These carbohydrates can be used directly by the plants or they can be transferred to other organisms which either feed on plant material or break down dead plant tissue. They can also be used for respiration. During respiration, chemical potential energy in glucose is transferred to another molecule, **ATP**. ATP is the immediate source of energy for various forms of biological work, such as movement, synthesis of large organic molecules and active transport.

Q1

A mouse is completely at rest in a comfortable environmental temperature. What are the main functions for which it needs ATP?

Energy and efficiency

We need to remember that none of these processes is totally efficient. During respiration, for example, not all the chemical potential from a molecule of glucose is transferred into molecules of ATP. Some of the energy in the glucose molecule will inevitably be lost as heat. This is obviously important when we come to look at the transfer of energy from one organism to another along a food chain or through a food web.

Figure 3 summarises the way in which energy is transferred within and between different organisms.

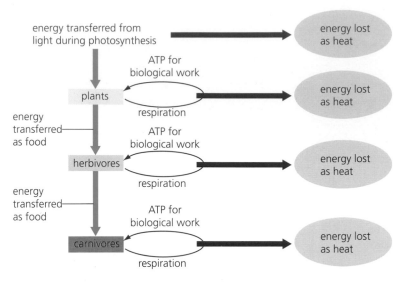

▲ **Figure 3** A summary of the way in which energy is transferred within and between organisms. Note that, each time a transfer occurs, some energy is lost as heat.

Photosynthesis

Animals must have a supply of complex organic molecules to provide the chemical potential energy they need to build new cells and tissues. These organic molecules come from food – sometimes from other animals but ultimately from plants. It is only by photosynthesis that light energy can be converted into chemical potential energy, and that simple inorganic molecules, such as carbon dioxide and water, can be built up into organic ones. Ultimately all animals rely on photosynthesis.

Photosynthesis is a complex process involving a number of separate reactions. It is useful to get an idea of the overall process before we look at any detail. There are two basic steps (Figure 4, opposite).

- **Light-dependent reactions:** light energy is captured by chlorophyll and some of this energy is transferred to chemical potential energy in ATP. A second substance is also produced. This is reduced **NADP**. In order to produce these substances, a molecule of water is split and oxygen is given off as a waste product.
- **Light-independent reactions:** ATP and reduced NADP are used in the conversion of carbon dioxide to carbohydrate.

Chloroplasts, chlorophyll and light energy

Structure of chloroplasts

You should remember from your AS course that chloroplasts are the site of photosynthesis. Each chloroplast (Figure 5) is surrounded by two plasma membranes. A system of membranes is also found inside the organelle. These membranes form flattened sacs called **thylakoids**. In some places the thylakoids are arranged in stacks called **grana**. More membranes join the grana to each other. The way in which the membranes are arranged is shown in Figure 6. The membranes that form the grana provide a very large surface area for chlorophyll molecules and other light-absorbing pigments.

Q2

Very little or no light penetrates to the ocean depths, yet some animals live there. Suggest how these animals rely on photosynthesis.

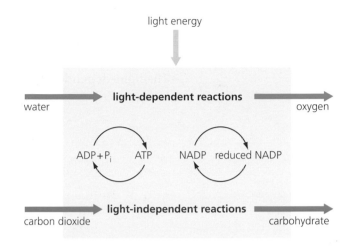

$$\text{carbon dioxide} + \text{water} \xrightarrow{\text{light energy}} \text{carbohydrate} + \text{oxygen}$$

▲ **Figure 4** The main steps in photosynthesis. The substances entering and leaving the leaf can be arranged to give the basic equation for photosynthesis.

These pigments form clusters called **photosystems**. The whole photosystem acts as a light-collecting system. Light energy is captured and passed from one pigment molecule to another before it finally reaches a chlorophyll molecule.

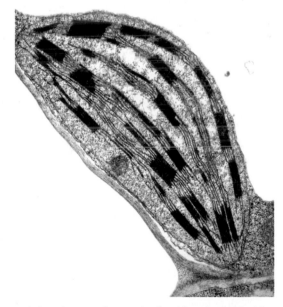

▲ **Figure 5** A transmission electron micrograph of a chloroplast. This chloroplast is approximately 5 μm in diameter.

The light-dependent reactions

If we crush up some nettle leaves in an organic solvent such as ethanol, we can make a chlorophyll solution. If we shine a bright light on this solution, it fluoresces. Instead of appearing green, it looks red. The light energy falling on the solution raises the energy level of some of the electrons in the chlorophyll. These electrons leave the chlorophyll molecule. They lose most of this energy as light of a different wavelength as they fall back into their places in the chlorophyll molecules. In a chloroplast, however, these

A simple model of the way in which the membranes inside a chloroplast are arranged. You will need some coins – about ten 10p pieces will do – and about three pieces of paper cut out like this:

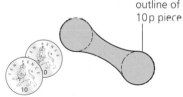

outline of 10p piece

Now arrange your coins in stacks. Vary the number of coins in each stack. Link them to each other with the pieces of paper you have cut out, like this:

paper linking coin stacks represents the membrane between grana

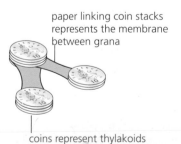

coins represent thylakoids arranged in stacks to form grana

▲ **Figure 6** Making a simple model to show how the membranes inside a chloroplast are arranged.

electrons do not return to the chlorophyll molecule from where they came. They pass down a series of electron carriers, losing energy as they go. This energy is used to produce ATP and reduced NADP.

The **light-dependent reactions** are described below and are summarised in Figure 7.

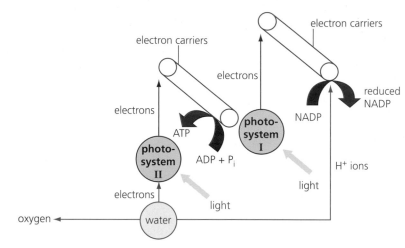

▲ **Figure 7** A summary of the light-dependent reactions of photosynthesis.

- Light strikes a chlorophyll molecule in **photosystem II**. The energy levels of two of the electrons in this chlorophyll molecule are raised. The electrons leave the chlorophyll molecule and pass to an **electron carrier**.
- The electrons are transferred along a series of electron acceptors which form an **electron transfer chain** in the chloroplast membranes. The electrons lose energy as they are passed from one electron carrier to the next. This energy is used to produce ATP.
- **Photolysis** takes place. This is the breakdown of a water molecule to produce protons, electrons and oxygen.

$$H_2O \rightarrow 2H^+ + 2e^- + \tfrac{1}{2}O_2$$
$$\text{water} \quad \text{protons} \quad \text{electrons} \quad \text{oxygen}$$

The electrons replace those lost from the chlorophyll molecule in photosystem I and oxygen is given off as a waste product. We will return to the protons later and see what happens to them.

- Light also strikes a chlorophyll molecule in **photosystem I**. Electrons leave this chlorophyll molecule and pass to an electron acceptor. They pass down another series of electron carriers. They are used, together with the hydrogen ions from the photolysis of water, to produce reduced NADP.

$$NADP + 2H^+ + 2e^- \rightarrow \text{reduced NADP}$$

Investigating the light-independent reactions

Calvin was a scientist from the University of California. He was the first person to sort out the sequence of the light-independent reactions in photosynthesis. He worked with single-celled photosynthetic organisms called algae. He grew the algae in the apparatus shown in Figure 8. He

$Q3$

In the light-dependent reactions of photosynthesis, what happens to the electrons that come from:
a) a water molecule
b) a chlorophyll molecule in photosystem I
c) a chlorophyll molecule in photosystem II?

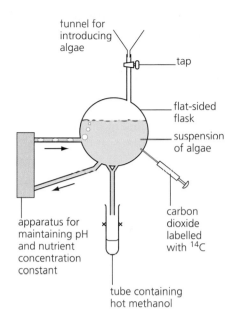

▲ **Figure 8** The apparatus Calvin used to investigate the light-independent reactions in photosynthesis.

added carbon dioxide labelled with a radioactive isotope of carbon, ^{14}C, to the contents of the flask. He collected samples of the algae by opening a tap and letting some of the contents of the flask drop into a tube containing hot methanol.

■ **In Calvin's investigation what was the purpose of the hot methanol?**

Hot methanol kills the algae immediately. This stops any further chemical reactions from taking place.

■ **Suggest why Calvin added carbon dioxide that had the radioactive isotope of carbon to the algae in the flask.**

Calvin wanted to find out what happened to the carbon dioxide. The radioactive carbon isotope acted as a label. He used a technique called autoradiography to identify those substances that contained ^{14}C. His technique involved using two-way paper chromatography to separate the substances extracted from the algal cells (Figure 9). The chromatogram was then dried and left in contact with a sheet of X-ray film in a dark room for several days. When developed, dark areas on the X-ray film showed the positions of substances containing ^{14}C.

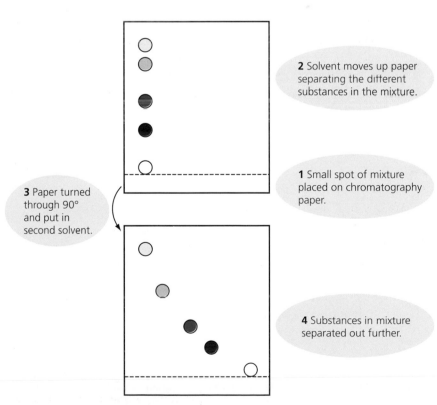

2 Solvent moves up paper separating the different substances in the mixture.

1 Small spot of mixture placed on chromatography paper.

3 Paper turned through 90° and put in second solvent.

4 Substances in mixture separated out further.

▲ **Figure 9** Two-way paper chromatography separates different substances in a mixture.

■ Calvin kept some algae in the flat-sided flask in the dark for some hours. He then added carbon dioxide containing ^{14}C. When he analysed the substances produced he did not find any that contained ^{14}C. What can you conclude from this?

The chemical reactions involving carbon dioxide do not take place in the dark. This could be because light is needed for the first reaction in the sequence or it could be that light produces some other substance that is needed in this reaction.

■ Calvin now left the algae in the light. He then switched the light off and immediately added carbon dioxide containing ^{14}C. When he analysed the substances present this time, he found labelled glycerate 3-phosphate (GP). What can you conclude from this?

This suggests that glycerate 3-phosphate is the first substance formed in the light-independent reactions. Clearly, the carbon dioxide must combine with another substance.

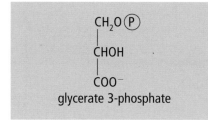

glycerate 3-phosphate

Look at the formula for this substance, on the left. You can see that it contains three carbon atoms and one phosphate group. This gave Calvin a clue. He suggested that carbon dioxide combined with ribulose bisphosphate (RuBP). RuBP contains five carbon atoms and two phosphate groups, so he predicted that the following reaction took place:

ribulose bisphosphate + carbon dioxide → 2 molecules of glycerate 3-phosphate

　　　5 carbons　　　　　　　1 carbon　　　　　　　　2 × 3 carbons

From his observations, Calvin predicted that something made in the presence of light was essential before this reaction could take place.

Now look at the graph in Figure 10. It shows how the amount of radioactive GP and RuBP changed when Calvin incubated algae with carbon dioxide labelled with ^{14}C.

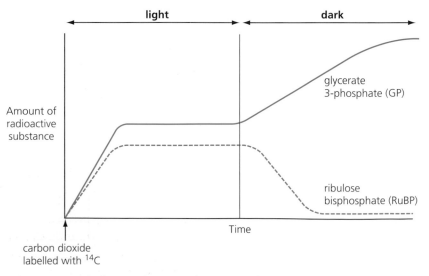

▲ **Figure 10** Graph showing changes in the amount of some radioactive substances in algae when they were given carbon dioxide labelled with ^{14}C.

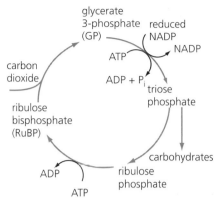

▲ **Figure 11** The light-independent reactions of photosynthesis form a cycle of reactions called the Calvin cycle. The diagram summarises the main steps in this cycle.

Q_4

Look at Figure 10. Use the information about the Calvin cycle to explain why the amount of RuBP falls when the light is switched off.

■ **Suggest an explanation for what happened to the amount of radioactive GP in the light.**

It rose at the start as carbon dioxide combined with RuBP to produce GP. It then levelled out probably because, although more GP was still being made, some was being converted to another substance.

■ **Why did the amount of GP increase in the dark?**

The most likely explanation is that light was necessary to convert GP into another substance. Without light, GP built up.

We will come back to the changes that occur in the amount of radioactive RuBP once we have looked at the rest of the biochemical pathway, called the **Calvin cycle**, that makes up the light-independent reactions (Figure 11). The main steps in the cycle are as follows.

• Carbon dioxide is accepted by RuBP to form two molecules of GP.
• GP is now converted to triose phosphate. This is a reduction reaction and requires two substances formed during the light-dependent reactions – reduced NADP and ATP.
• Triose phosphate is a carbohydrate. Some of it is built up into other carbohydrates, such as glucose and starch, and into amino acids and lipids. The rest is used to make ribulose phosphate.
• Ribulose phosphate is finally converted to RuBP using phosphate from ATP.

Limiting factors and photosynthesis

Plants rely on photosynthesis to produce the carbohydrates and other organic substances they need for growth. The greater a plant's rate of photosynthesis, the greater its rate of growth and, if we are considering crop plants, the higher the yield. Among the environmental factors that affect the rate of photosynthesis are:

• light intensity
• carbon dioxide concentration
• temperature.

Looking at complex data

The graph in Figure 12 on the next page shows the effects of light intensity, carbon dioxide concentration and temperature on the rate of photosynthesis.

■ **What does curve A tell you about the effect of light intensity on the rate of photosynthesis?**

Over the first part of the curve, the rate of photosynthesis is directly proportional to light intensity. Because of this, we can describe light as limiting the rate of photosynthesis over this part of the curve. As light intensity continues to increase, the curve starts to flatten out – light intensity is no longer limiting the rate of photosynthesis.

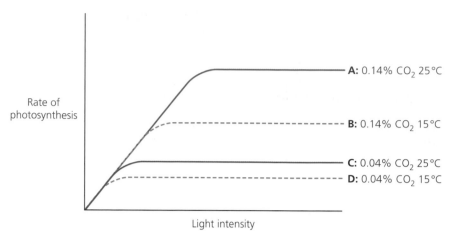

A: 0.14% CO_2 25°C

B: 0.14% CO_2 15°C

C: 0.04% CO_2 25°C

D: 0.04% CO_2 15°C

Rate of photosynthesis

Light intensity

▲ **Figure 12** The effects of light intensity, carbon dioxide concentration and temperature on the rate of photosynthesis.

■ **At low light intensities, such as on a winter's day in the UK, what factor limits the rate of photosynthesis?**

This is most likely to be light intensity. Increasing carbon dioxide concentration or temperature will not increase the rate of photosynthesis when light intensity is low.

■ **In bright conditions, what factor is most likely to be limiting the rate of photosynthesis?**

Carbon dioxide limits the rate of photosynthesis in bright conditions. Increasing the concentration of carbon dioxide in the atmosphere from its normal level of approximately 0.04% to 0.14% and keeping the temperature constant has a much greater effect on the rate of photosynthesis than increasing the temperature from 15°C to 25°C and keeping the carbon dioxide concentration constant.

■ **Under what environmental conditions do you think it would be worth heating a glasshouse to increase the yield of a crop growing in it?**

You can see from the graph that increasing the temperature from 15°C to 25°C increases the rate of photosynthesis. The effect is greatest when neither light nor carbon dioxide concentration are limiting.

Commercial glasshouses (Figure 13) can use the methods listed below to control the environmental factors that normally limit the rate of photosynthesis and thus crop yield can be increased.

* Artificial lighting can be used if the intensity of natural light falls to a level where it limits the rate of photosynthesis. Very high light intensities damage chlorophyll and therefore limit the rate of photosynthesis. Automatic blinds can be used to shade plants under these conditions.
* The concentration of carbon dioxide can be increased either by pumping the gas in or by using paraffin heaters. Tomato plants grow best when the carbon dioxide concentration is maintained at 0.1%.

▲ **Figure 13** Food crops such as tomatoes are grown commercially in glasshouses where environmental factors can be controlled resulting in high yields.

• The greenhouse can be heated. This prevents night temperatures falling to levels that might damage the plants but also ensures optimum temperature conditions exist.

Carbon dioxide concentration and crop production

We can deliberately increase the carbon dioxide concentration inside a glasshouse but how will the rising atmospheric carbon dioxide concentration affect the growth of field crops? In bright conditions, the concentration of carbon dioxide usually limits the rate of photosynthesis. An increase in carbon dioxide concentration should therefore increase the rate of photosynthesis. Look at Figure 14. It shows the results of a laboratory experiment in which scientists investigated the effect of an increase in carbon dioxide concentration on the rate of photosynthesis of wheat and maize plants.

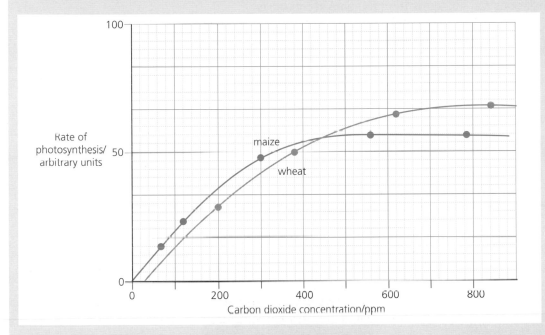

▲ **Figure 14** The effect of carbon dioxide concentration on the rate of photosynthesis in wheat and maize plants.

Wheat is a **C₃ plant**. This means it photosynthesises using the biochemical pathway described on page 48. In this pathway, carbon dioxide combines with RuBP to produce a substance containing three carbon atoms. Many of the important food plants grown in temperate regions, such as sugar beet and potatoes, are also C₃ plants. They respond to an increase in carbon dioxide concentration in a similar way to wheat.

1 Suppose we doubled the concentration of carbon dioxide from 330 ppm to 660 ppm. What would happen to the rate of photosynthesis in wheat?

2 Can you conclude from this graph that an increase in the concentration of carbon dioxide in the atmosphere will result in an increase in the growth of wheat plants? Explain your answer.

3 Will an increase in carbon dioxide affect the rate of photosynthesis of all crop plants in the same way?

Look at the data in Table 1, overleaf. The data in this table refer to C₃ plants and predict the effect of doubling current atmospheric concentrations of carbon dioxide.

2 Energy transfer

Plant	Percentage increase in biomass	Percentage increase in marketable yield
Cotton	124	104
Tomato	40	21
Cabbage	37	19
Weeds	34	not applicable

Table 1 The effect of doubling current atmospheric concentrations of carbon dioxide on some C_3 plants.

4 What general conclusions can you draw from the data in this table?

5 Use the data in Table 1 to suggest how doubling the current atmospheric concentration of carbon dioxide would affect the marketable yield of cabbages growing in a field.

How efficient is photosynthesis?

A lot of light energy falls on the surface of the Earth. Only a small part of this is converted into chemical potential energy during photosynthesis. Look at the maize crop shown in Figure 15.

energy from sunlight

energy in light reflected: 15%

energy lost as heat: 50%

energy in light passing through leaves and not striking chlorophyll: 30%

▲ **Figure 15** This maize crop converts only a very small percentage of the total energy in visible light into chemical potential energy in organic substances.

- About 50% of the visible-light energy falling on the plants in this crop is absorbed and converted into heat energy. Much of this heat energy evaporates water from the surface of the soil and from leaves during transpiration.

- Approximately 15% of the visible-light energy is reflected.
- Almost a third of the visible-light energy, approximately 30%, is transmitted. It passes directly through the plants without striking any chlorophyll molecules.

You can see from this that only a very small percentage of the total visible-light energy can be converted to chemical potential energy in substances produced by photosynthesis.

The rate at which a plant is able to produce organic substances as a result of photosynthesis is called its **gross productivity**. We can measure this as an increase in either:

- dry mass, measured in units such as $g\,m^{-2}\,day^{-1}$
- or chemical potential energy in the organic substances, measured in units such as $kJ\,m^{-2}\,day^{-1}$.

Some of the substances formed during photosynthesis are not, however, used to form new cells and tissues. They are used in respiration. The difference between gross productivity and respiration is **net productivity**. We can calculate net productivity from the equation:

$$\text{net productivity} = \text{gross productivity} - \text{respiration}$$

Net productivity is important because it represents the amount of food available to organisms in the next trophic level in a food web. It is also important to agricultural scientists because it provides them with an indication of the potential yield of a crop.

Calculating the efficiency of photosynthesis

Figure 16 shows heather plants. Scientists have found heather particularly useful in studying the efficiency of photosynthesis because:

- In moorland areas, heather grows in large clumps called stands. Each stand is almost pure heather so we do not have to consider what fraction of the total visible-light energy is used by other species.
- In many moorland areas, heather is managed. This means that it is burnt at regular intervals. Old woody plants are replaced by young plants which provide a better food supply for game birds, such as grouse. Estate managers keep records so we usually know the age of a particular stand of heather.
- Many investigations have been carried out using heather. By sharing the findings of their research, scientists have been able to replicate and further test their work. This increases the reliability of the conclusions that they draw.

We can calculate the efficiency of photosynthesis in heather from the formula:

$$\text{efficiency of photosynthesis} = \frac{\text{chemical potential energy in heather plants}}{\text{total amount of light energy falling on heather plants}}$$

To calculate the figure on the top line of the equation we need to multiply the chemical potential energy in 1 g of heather by the biomass of heather produced in $g\,m^{-2}\,year^{-1}$.

▲ **Figure 16** Heather.

Finding the chemical potential energy in 1 g of heather

We can find this by burning a sample of heather in a food-fuel calorimeter (Figure 17) and measuring the energy released as heat.

Look at Figure 17 and answer the questions.

1 An oxygen supply is connected to the apparatus. What is the advantage of burning the heather in oxygen and not in air?

2 A copper spiral is attached to the top of the combustion chamber. What is the function of this copper spiral?

3 The water jacket contains a large volume of water. The total rise in temperature of the water in this jacket when the heather is burnt is only small. Explain the advantage of a small rise in temperature.

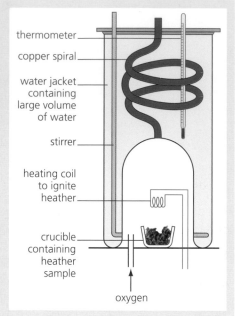

▶ **Figure 17** A food-fuel calorimeter used to measure the energy released when a sample of heather is burnt.

Table 2 shows some typical results obtained from the apparatus in Figure 17.

Volume of water in water jacket/cm³	650
Temperature of water before the heather sample was burnt/°C	18
Temperature of water after the heather sample was burnt/°C	23
Mass of heather/g	0.5

Table 2 Results obtained from burning a sample of heather in a food-fuel calorimeter.

■ The amount of energy needed to raise the temperature of 1 cm³ of water by 1 °C is 4.2 joules. Calculate the amount of energy released by burning 1 g of heather.

In this investigation, the energy released has raised the temperature of 650 cm³ of water by 5 °C. So, the total amount of energy released by the heather sample is $4.2 \times 650 \times 5$ J.

This is the amount of energy released by 0.5 g of heather. We need to divide the total amount of energy released by 0.5 to get the amount of energy released per gram:

$$\frac{4.2 \times 650 \times 5 \, \text{J} \, \text{g}^{-1}}{0.5}$$

This gives us a figure of $27\,300 \, \text{J} \, \text{g}^{-1}$ or $27.3 \, \text{kJ} \, \text{g}^{-1}$.

■ Do you think that the figure that we have calculated is an overestimate or an underestimate? Explain your answer.

It is probably an underestimate. There are several reasons for this. Not all the heather will have been burnt. The ash that is left in the crucible may contain some chemical potential energy that has not been released. In addition, it is unlikely that all the heat will have been transferred to the water in the water jacket.

Finding the biomass of heather produced

The graph in Figure 18 shows the mean biomass of heather from plants of different ages growing on an area of moorland in Yorkshire.

1 Suggest how quadrats could have been used to collect the heather samples used to provide this data.

2 The biomass of the heather is given as dry biomass.

  a) Describe how you would measure the dry biomass of a heather sample.

  b) How would you make sure that your value for dry biomass was reliable?

  c) What is the advantage of measuring biomass as *dry* biomass?

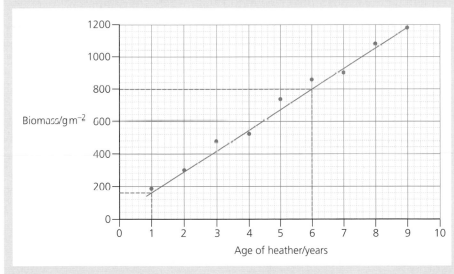

▲ **Figure 18** The mean dry biomass of heather plants of different ages.

■ Explain how you would use the graph in Figure 18 to calculate the mean annual increase in dry biomass of the heather plants.

A curve of best fit has been drawn through the points on the graph. This curve is a straight line. The mean annual increase in dry biomass can be calculated by finding the gradient of this line. The construction lines on the graph show that the biomass increases by 640 g in five years. This is an increase of 128 g year^{-1}.

■ What is the evidence from the graph that the heather did not increase in biomass by the same amount each year?

The points do not all lie on the curve of best fit.

■ Suggest why the heather did not increase in biomass by the same amount each year.

Plant growth does not only depend on the amount of light falling on the leaves. Other abiotic factors, such as rainfall and temperature, also affect growth. Biotic factors, such as grazing by large mammals (for example, deer and sheep) or by smaller organisms (for example, caterpillars of the emperor moth), also affect biomass.

■ The figure that we have calculated underestimates the mean annual increase in the total biomass of the heather plants. Suggest why.

It is very likely that the heather samples were only the parts that grew above ground. Root growth was probably not considered, nor was the biomass of dead flowers and the seeds that had fallen from the plants.

Scientists estimated the total amount of light energy falling on one square metre of the heather moorland in Yorkshire where the samples were collected to be 1 415 000 kJ. They based this estimate on the amount of light falling on the heather in the growing season so it does not include the winter months when the temperatures are too low for growth.

■ Use all the figures in this section to calculate the efficiency of photosynthesis in these heather plants.

We will need to return to the original equation:

$$\text{efficiency of photosynthesis} = \frac{\text{chemical potential energy in heather plants}}{\text{total amount of light energy falling on heather plants}}$$

In more detail this is:

$$\text{efficiency} = \frac{\text{chemical potential energy in 1 g of heather} \times \text{biomass of heather}}{\text{total amount of light falling on 1 m}^2 \text{ of heather moorland}} \times 100\%$$

$$= \frac{27.3 \times 128}{1\,415\,000} \times 100\%$$

$$= 0.2\%$$

Crops and photosynthetic efficiency

Heather is not a particularly productive plant. Heather moorland is often associated with upland areas where many abiotic factors limit growth. Temperatures are often low and the acid soils in which heather grows are often poor in essential nutrients, such as nitrates.

Crop plants growing in the UK are a little more efficient and are able to convert approximately 2% of the light energy falling on them into chemical potential energy. A tropical crop, however, such as the sugar cane in Figure 19, is able to convert between 7 and 8% of the light energy it receives into chemical potential energy. It is probably the most productive of all plants. Why is it so efficient?

- Light intensity is in general much higher in the tropics than in temperate regions. During the daylight hours it is unlikely to limit the rate of photosynthesis. Tropical temperatures are also higher than those in temperate regions. The rates of enzyme-controlled reactions involved with synthesis and growth are therefore likely to be faster in a tropical crop.
- Sugar cane is a grass. It therefore has tall upright leaves. This structure is very efficient at intercepting light.
- Sugar cane is a cultivated plant grown in plantations. It is artificially irrigated and is provided with fertiliser, so lack of water and mineral ions are unlikely to limit its growth. In addition, weeds are controlled so the light energy goes to produce chemical potential energy in the sugar cane plants, not in other plants.

▲ **Figure 19** Sugar cane is a crop plant and, like all crop plants, it has been selected over many generations for particular characteristics. One of these is the efficiency with which it converts light energy into chemical potential energy in useful substances.

Respiration

What is it all about?

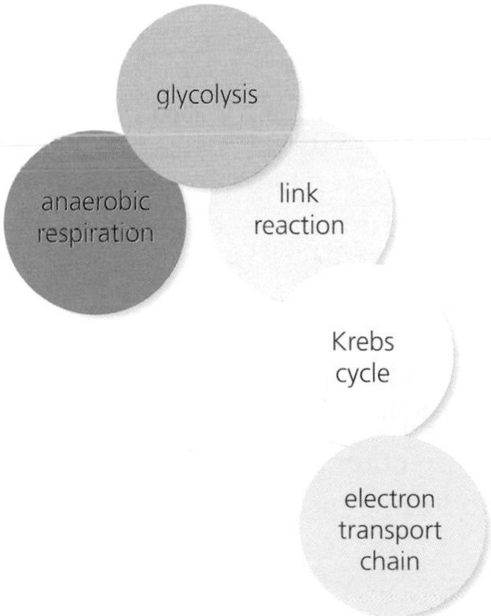

Respiration takes place in all living cells. It is a biochemical process in which organic substances are used as fuel. They are broken down in a series of stages and the chemical potential energy they contain is transferred to ATP. We can summarise the process with the equation:

$$C_6H_{12}O_6 + 6O_2 \rightarrow 6CO_2 + 6H_2O + energy$$

Unfortunately this equation is misleading in a number of ways. It shows the fuel as glucose. In many living cells the main fuel is glucose, but fatty acids, glycerol and amino acids are also **respiratory substrates** and can be used for respiration. The equation also shows that oxygen is required – this is correct when aerobic respiration is occurring but respiration can, however, also take place anaerobically. **Anaerobic respiration** means respiration without oxygen. Finally, the equation shows respiration as a single reaction. It isn't. It involves a number of reactions in which the respiratory substrate is broken down in a series of steps, releasing a small amount of energy each time. The steps involved in the respiration of glucose are summarised in Figure 20.

glycolysis

anaerobic respiration

link reaction

Krebs cycle

electron transport chain

▲ **Figure 20** The breakdown of glucose in respiration. Aerobic respiration takes place in the presence of oxygen. Respiration can continue when there is no oxygen. This is anaerobic respiration.

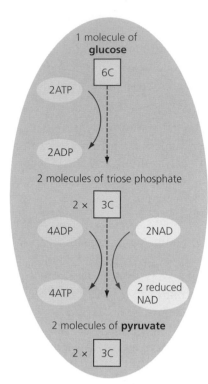

▲ **Figure 21** A summary of glycolysis – the first step in respiration. The boxes show the number of carbon atoms in the different molecules. There is a net gain of two molecules of ATP from each molecule of glucose.

Q_5

Suggest why the pathway from glucose to pyruvate in Figure 21 is shown by broken arrows.

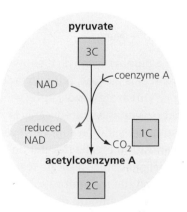

▲ **Figure 22** The link reaction. The boxes show the number of carbon atoms in the molecules involved in the respiratory pathway.

Glycolysis

Glycolysis (Figure 21) is the first step in the biochemical pathway of respiration.

- Glucose contains a lot of chemical potential energy. In order to release this energy, we need to use some energy from ATP. In the first stage of glycolysis, a molecule of glucose is converted into two molecules of triose phosphate. This requires two molecules of ATP.
- Each molecule of triose phosphate is then converted to pyruvate. This process produces ATP. A total of four molecules of ATP are produced, two for each triose phosphate molecule. During glycolysis then, there is a net gain of two molecules of ATP for each molecule of glucose.
- The conversion of triose phosphate to pyruvate is an oxidation reaction and involves the reduction of a coenzyme called NAD. NAD is converted to reduced NAD as a result.

NAD, oxidation and reduction

Oxidation is sometimes represented as the addition of oxygen to a substance but, more accurately, it is any reaction in which electrons are *removed*. **Reduction**, on the other hand, involves the gain of electrons. Whenever one substance is oxidised, another must be reduced. In simple terms, if one substance loses electrons, another must gain them. We often use the term **redox reaction** for a reaction in which one substance is oxidised and another is reduced.

In glycolysis, the conversion of triose phosphate to pyruvate is an oxidation. Therefore electrons are lost. NAD gains these electrons, so it is reduced and becomes **reduced NAD**.

The link reaction

Pyruvate still contains a lot of chemical potential energy. When oxygen is available, this energy can be made available in a series of reactions known as the Krebs cycle. The **link reaction** is a term used to describe the reaction linking glycolysis and the Krebs cycle. It is summarised in Figure 22.

- Pyruvate combines with coenzyme A to produce acetylcoenzyme A.
- Pyruvate contains three carbon atoms. Acetylcoenzyme A contains several carbon atoms but only two of these enter the Krebs cycle. One of the carbon atoms from pyruvate goes to form a molecule of carbon dioxide.
- The conversion of pyruvate to acetylcoenzyme A is an oxidation reaction and, like glycolysis, also involves the reduction of NAD.

Q_6

Copy and complete this equation using information from Figure 22.

pyruvate + coenzyme A + NAD →

Working out what happens in the Krebs cycle

The **Krebs cycle** (Figure 23) is a cycle of reactions. Look at the figure carefully and use it to answer these questions.

1 There are four carbon atoms in oxaloacetate. Explaining your answer in each case, how many carbon atoms are there in:

   a) citrate

   b) α-ketoglutarate

   c) succinate?

2 The reactions in which citrate is converted via a series of intermediate substances to oxaloacetate are oxidation reactions.
   What is the evidence from the cycle in Figure 23 that these are oxidation reactions?

3 The Krebs cycle produces reduced NAD. How many molecules of reduced NAD are produced in the reactions that make up one complete cycle?

4 How many molecules of reduced NAD are produced in the reactions that make up the Krebs cycle for each glucose molecule?

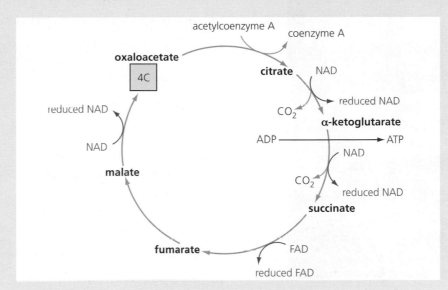

▲ **Figure 23** Some of the reactions that make up the Krebs cycle.

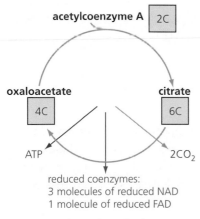

▲ **Figure 24** The Krebs cycle plays a very important part in respiration. It is the main source of reduced coenzymes, which are used to produce ATP in the electron transport chain.

Figure 24 is a simple diagram which summarises the essential features of the Krebs cycle.

- Acetylcoenzyme A, produced in the link reaction, is fed into the cycle. It combines with a 4-carbon compound (oxaloacetate) to produce a 6-carbon compound (citrate).
- In a series of reactions, the citrate is converted back to oxaloacetate. In this process two molecules of carbon dioxide are given off.
- For each complete cycle, one molecule of ATP is produced.
- The Krebs cycle reactions involve oxidation. This process involves the reduction of coenzymes. Each complete cycle produces three molecules of reduced NAD and one molecule of another reduced coenzyme, reduced FAD. The most important function of the Krebs cycle in respiration is the production of reduced coenzymes. They are passed to the **electron transport chain**, where the chemical potential energy that these molecules contain is used to produce ATP.

Mitochondria and the electron transport chain

Reduced coenzymes have so far been produced at each step in the respiratory pathway:

- reduced NAD is produced in glycolysis, the link reaction and the Krebs cycle
- reduced FAD is produced in the Krebs cycle.

2 Energy transfer

▲ **Figure 26** When a cuckoo pint flowers, the temperature of the spadix increases until it is as much as 15 °C above the temperature of the environment. This increase in temperature causes molecules of a substance similar to substances found in faeces and decaying bodies to be released. This attracts the flies that pollinate cuckoo pint flowers.

Q7

The cells in the spadix of a flowering cuckoo pint (Figure 26) respire. When electrons from reduced NAD pass down the electron transport chain, ATP is not produced. Use this information to explain the increase in temperature of the spadix in a flowering cuckoo pint.

Q8

The chemical potential energy in a molecule of glucose can be used to produce 38 molecules of ATP in aerobic respiration. How many molecules of ATP can one molecule of glucose produce in anaerobic respiration?

We will start by looking at what happens to the reduced NAD. It passes to chains of protein molecules situated on the internal membranes of **mitochondria**. These proteins act as **electron carriers** and form the electron transport chain.

Figure 25 is a simplified diagram showing how electrons from reduced NAD pass from one protein to the next along this electron transport chain. To avoid unnecessary detail, we have called these proteins carrier 2, carrier 3 and so on.

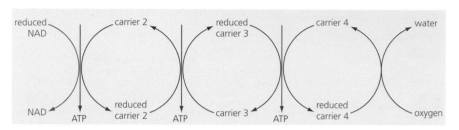

▲ **Figure 25** Energy is released as electrons pass from one carrier to the next in the electron transport chain. Some of this energy is lost as heat but a lot of it goes to produce ATP. Each molecule of reduced NAD entering the chain produces three molecules of ATP. A molecule of reduced FAD produces two molecules of ATP.

The electrons pass from the reduced NAD to carrier 2. The reduced NAD is oxidised because it has *lost* electrons and carrier 2 is reduced because it has *gained* electrons. When these electrons are transferred, a small amount of energy is released. Some of this energy is lost as heat but some is used to pump hydrogen ions out through the inner mitochondrial membrane into the space between the inner membrane and the outer membrane. There is now a greater concentration of hydrogen ions in the space between the membranes than there is in the matrix of the mitochondrion. The only way in which the hydrogen can get back through the inner membrane into the matrix is through ATPase enzymes. As they pass through these enzymes, enough potential energy is released to form ATP.

The last molecule in the chain is oxygen. Oxygen combines with hydrogen ions and electrons to produce water.

Anaerobic respiration

Sometimes there is not enough oxygen for an organism to respire aerobically using the pathways we have just described. Under these conditions, it produces ATP by respiring anaerobically. The only stage in the anaerobic pathway which produces ATP is glycolysis, so the process is not as efficient as aerobic respiration.

Look back at Figure 21 (page 56). You will see that during glycolysis NAD is reduced. Reduced NAD is normally converted back to NAD when its electrons are passed to the electron carriers in the electron transport chain. This can only happen when oxygen is present. Obviously, if all the NAD in a cell was converted to reduced NAD, the process of respiration would stop.

In anaerobic respiration in animals pyruvate is converted to **lactate**. In plants and microorganisms, such as yeast, it is converted to **ethanol** and **carbon dioxide**. In both of these pathways (Figure 27), reduced NAD is converted back to NAD. This allows glycolysis to continue.

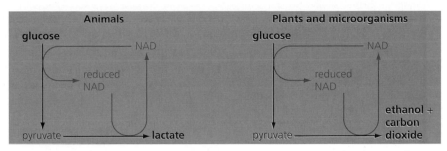

▲ **Figure 27** Anaerobic respiration allows organisms to produce ATP in the absence of oxygen.

Energy transfer

On page 43 we saw that photosynthesis was the main route by which energy enters an ecosystem. There are other ways, however, by which it can enter. Let us consider the bottom of the Pacific Ocean. It is pitch dark. No light penetrates its depths. It is also cold. The water remains just above freezing all year round. In a few areas volcanic vents bubble out a mixture of sulphur-rich gases. You might think that life could not possibly exist in these conditions, but it can!

Around the volcanic vents, bacteria are found. These bacteria use the sulphur-containing substances bubbling from the vents. They obtain energy from chemical reactions involving these substances and use it to synthesise the complex organic molecules that make up their cells. The bacteria support large worms and other invertebrate animals. A community of living organisms exists because chemical potential energy in large molecules is transferred from one organism to another through a food web.

In whatever way energy enters an ecosystem, it is transferred through food chains and food webs. Ecologists study feeding habits of organisms so that they can investigate food webs. Some species can be watched directly. The larvae of butterflies and moths – caterpillars – can be observed feeding on the leaves of particular species of plants. The food of other species, such as the badger (Figure 28), can be identified by studying the remains in faeces.

Q9
Rice is grown in swampy conditions. The cells in rice roots are very tolerant to high concentrations of ethanol. Explain how this is an advantage to a rice plant.

Q10
Organisms that use light energy to produce organic substances from simple inorganic substances are called photoautotrophs. What is a chemoautotroph?

▲ **Figure 28** Badgers are large conspicuous animals but they are nocturnal and difficult to watch feeding. Badger faeces are easily recognisable and contain undigested remains, such as these plum stones.

With other animals, particularly small invertebrates, it is very difficult to make direct observations of feeding behaviour. You would be very lucky to see, for example, a ground beetle catch and eat its prey. Even if you did see such an event, you would not know whether it was normal feeding behaviour or whether it was unusual. Ecologists sometimes make use of antibodies to provide information about feeding behaviour.

Using antibodies to investigate feeding

Ecologists can dissect some animals and identify the food remains in their guts. But suppose they wanted to know if an animal, such as a ground beetle, fed on slugs. There wouldn't be any hard parts to identify in the beetle's gut. Some of the proteins that made up the slug's body, however, would be present. Ecologists can use an **enzyme-linked immunosorbent assay (ELISA)** to confirm that these proteins come from a specific organism. Figure 29 shows how an ELISA test is used to find out whether ground beetles eat slugs.

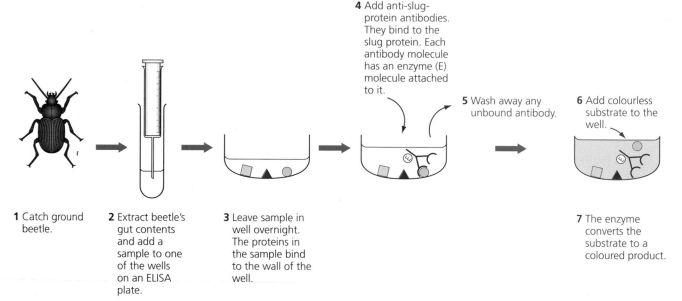

4 Add anti-slug-protein antibodies. They bind to the slug protein. Each antibody molecule has an enzyme (E) molecule attached to it.

5 Wash away any unbound antibody.

6 Add colourless substrate to the well.

1 Catch ground beetle.

2 Extract beetle's gut contents and add a sample to one of the wells on an ELISA plate.

3 Leave sample in well overnight. The proteins in the sample bind to the wall of the well.

7 The enzyme converts the substrate to a coloured product.

▲ **Figure 29** Using an ELISA test to confirm that ground beetles eat slugs.

Before we look at the diagram in detail, there are two important principles about which we must remind ourselves. We first met these principles in Unit 1.

- Different organisms contain different proteins. Suppose our ground beetle had been feeding on slugs and, say, earthworms. Some of the proteins found in the slugs would probably be very similar to proteins found in the earthworms so we probably couldn't tell which animal the protein concerned came from. Some, however, would be different. These specific slug proteins would have specific sequences of amino acids and their molecules would have specific tertiary structures.
- Antibodies are also proteins. They have specific binding sites. These binding sites mean that they will only bind to molecules that have a complementary shape. An anti-slug-protein antibody will only bind to one particular slug protein. It won't bind to proteins from any other animal unless they are identical to the slug protein.

Steps 1 and 2 in Figure 29 should be easy enough to understand but we may need to explain some of the other steps. We will start with step 3 where the sample of gut contents has been left overnight in one of the wells on the ELISA plate.

■ **Three protein molecules are shown attached to the wall of the well. Why do these protein molecules have different shapes?**

Each shape represents a different protein so there are three different proteins shown here. All three could be slug proteins, but it is possible that one or two of them might have come from other animals that the ground beetle had eaten.

■ **Step 4 shows anti-slug-protein antibody binding only to a slug protein. Why does this antibody bind only to slug protein?**

This is the point we made earlier. Antibodies are specific and an anti-slug-protein antibody will only bind to the protein shown as a blue circle. This is a slug protein.

■ **Why is the unbound antibody washed away (step 5)?**

If we don't wash the unbound antibody away, it will remain in the well. The enzyme on the unbound antibody will result in the coloured product being formed even if no slug protein is present.

■ **Not all ground beetles eat slugs. Explain how you would be able to tell if the ground beetle from which you had obtained the gut sample had not been eating slugs.**

There would be no slug proteins attached to the wall of the well to bind to the anti-slug-protein antibodies. Therefore there would be no enzyme to catalyse the reaction in which the colourless substrate was converted to a coloured product.

■ **How could you use an ELISA plate to find out whether slugs were important items of food for ground beetles?**

You could add samples from different ground beetle guts to different wells on the ELISA plate. By counting the number of wells where there was a colour change, you could find the percentage of ground beetles that ate slugs.

Food chains and food webs

In any ecosystem different organisms gain their nutrients in different ways (Figure 30). Green plants are **producers**. They produce organic substances from carbon dioxide and water. They rely on photosynthesis to transfer energy from sunlight to chemical potential energy in organic substances. The other organisms that make up the community rely either directly or indirectly on the organic substances produced by the producers. **Primary consumers** feed on producers. **Secondary consumers** feed on primary consumers and **tertiary consumers** feed on secondary consumers. Organisms that are not eaten eventually die. Another group of organisms, the **decomposers**, break down dead tissues and use the organic substances that make up these tissues as a source of chemical potential energy.

Q11

Sundew is an insectivorous plant. It has sticky hairs on its green leaves. These hairs trap small insects, such as aphids, which feed on plant sap. What trophic levels does the sundew plant occupy?

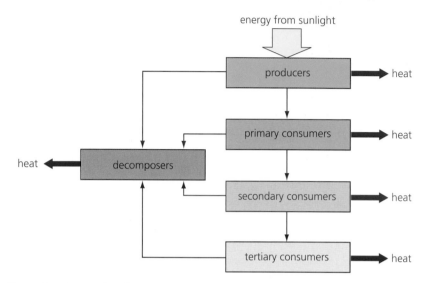

▲ **Figure 30** The transfer of energy in an ecosystem. The boxes represent trophic levels. The arrows show the direction in which energy is transferred.

We often talk about **food chains** suggesting perhaps that we frequently encounter situations where animal **B** feeds only on plant **A**. In turn, animal **C** only eats animal **B**, and animal **D** only eats animal **C**. This hardly ever happens under natural conditions. Food chains are linked to each other to form complex **food webs**. We will look at the patch of nettles that we showed in Figure 4, Chapter 1, page 3. Figure 31 shows a possible food web associated with a nettle patch.

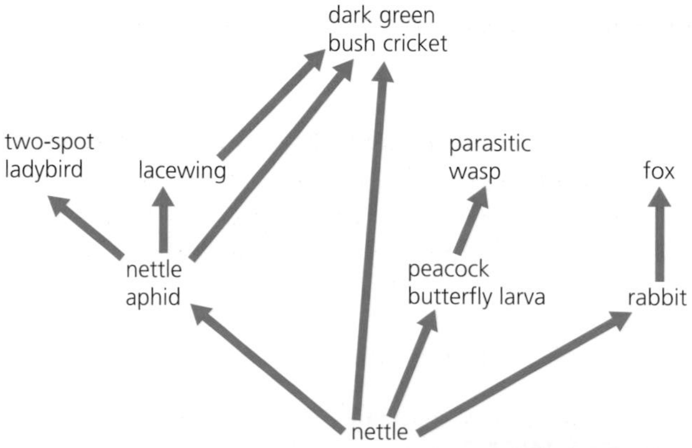

Note that:

- some organisms, such as the dark green bush cricket, feed at different trophic levels
- some organisms feed on different food when they are larvae and when they are adults. The larva of the peacock butterfly, for example, eats nettle leaves. The adult butterfly feeds on nectar produced by flowers.

▲ **Figure 31** A simplified food web showing some of the organisms that feed on nettles.

Food webs and human food

Table 3 shows the mean trophic levels of marine fish caught for human food in the period 1950 to 2000.

Year	Mean trophic level
1950	3.37
1960	3.36
1970	3.39
1980	3.29
1990	3.26
2000	3.21

Table 3 The mean trophic levels of marine fish caught for human food, 1950–2000.

1 a) Phytoplankton consists of single-celled photosynthetic organisms that float in the surface water. A species of fish feeds only on phytoplankton. What would be the trophic level of this species of fish?

b) Another species of fish has a trophic level of 3.0. Is this fish a primary consumer, a secondary consumer or a tertiary consumer?

2 a) Describe how the mean trophic level of the marine fish catch has changed over the period 1950 to 2000.

b) Suggest an explanation for the change you described in your answer to question 2 a).

3 Over the same period of time, more cattle and sheep have been fed on protein concentrates as well as grass. These protein concentrates are often made from animal material. Suggest how the mean trophic levels of farm animals have changed over the period shown in the table.

Ecological pyramids

The diagram in Figure 31 provides *qualitative* information. It only shows what food different organisms eat. Biologists are also interested in *quantitative* information. This is where ecological pyramids are useful.

Pyramids of numbers

The simplest way to compare the different trophic levels in an ecosystem is to count the number of organisms present. We can represent this information as a **pyramid of numbers**. Three typical pyramids of numbers are shown in Figure 32, overleaf. If you look at the pyramid representing the food chain

nettle → rabbit → fox

you will see that it is pyramid shaped. Obviously, if you were to count all the nettle plants in the area concerned, there would be a lot more than the number of rabbits. In turn, there would be more rabbits than foxes. There are two important exceptions to this pyramid shape when we are considering numbers of organisms. The pyramid will be upside-down or inverted if many small animals feed off a large plant. This is the case in the second example in Figure 32, where the primary consumers are aphids.

Another example of an inverted pyramid is where an organism is supporting a large number of small parasites, such as humans and head lice or, in the third example in Figure 32, parasitic wasps feeding inside a peacock butterfly larva.

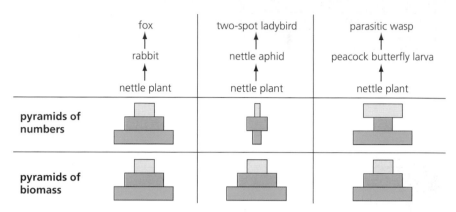

▲ **Figure 32** These pyramids all relate to food chains based on nettle plants.

Pyramids of biomass

Biomass is a measure of the total amount of living material present. If you were to shake all the aphids from a clump of nettles and weigh them, you would have the biomass of the aphids on that clump of nettles. A pyramid of biomass allows us to compare the mass of organisms at each trophic level of a particular food chain or food web. Pyramids of biomass overcome the problem of trying to compare numbers when the organisms are of very different size. Because of this, all pyramids of biomass shown in Figure 32 have a similar pyramid shape.

There is one important exception to this pyramid shape – organisms, such as the phytoplankton shown in Figure 33, which multiply very rapidly. At any one time, their total biomass may be less than the biomass of the primary consumers that feed on them.

Pyramids of energy

On page 51 we considered the efficiency with which energy is transferred to plants in photosynthesis. In this section we will look at the efficiency with which energy is transferred to the consumers in an ecosystem. Figure 34 shows the percentage of energy transferred between different trophic levels. We can look at this in a different way. If we take a figure of 2% as representing the percentage of light energy converted to chemical potential energy in plants, then for every 10 000 kJ of light energy absorbed by the producers, 200 kJ will be incorporated into their tissues. Similarly, if we assume that about 10% of the chemical potential energy in producers becomes chemical potential energy in primary consumers, then 20 kJ from our original 10 000 kJ will be transferred to the tissues of primary consumers. At each step less chemical potential energy will be transferred. We can construct a third type of ecological pyramid that allows us to compare the amount of energy transferred from one trophic level to the next in a given period of time. This is called a **pyramid of energy**. Pyramids of energy are always pyramid shaped. There are no exceptions.

▲ **Figure 33** Phytoplankton are tiny single-celled algae. They are producers and float in the surface waters of lakes and seas. When conditions are favourable, they multiply very rapidly. Because of this, their total biomass at any one time may be less than that of the primary consumers. However, if you were to look at their biomass over a period of, say, a month, then it would be greater than that of the primary consumers.

Q12

It is rare for there to be more than five trophic levels in a food web. Explain why.

▲ **Figure 35** Crocodiles use environmental temperature to help them to maintain their optimum body temperature.

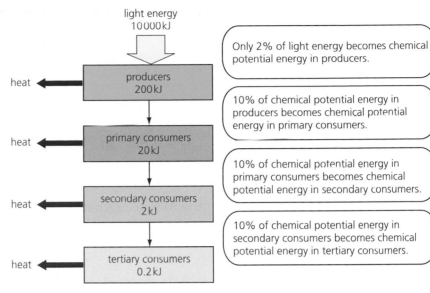

light energy
10 000 kJ

heat ◄— producers
200 kJ

heat ◄— primary consumers
20 kJ

heat ◄— secondary consumers
2 kJ

heat ◄— tertiary consumers
0.2 kJ

Only 2% of light energy becomes chemical potential energy in producers.

10% of chemical potential energy in producers becomes chemical potential energy in primary consumers.

10% of chemical potential energy in primary consumers becomes chemical potential energy in secondary consumers.

10% of chemical potential energy in secondary consumers becomes chemical potential energy in tertiary consumers.

▲ **Figure 34** Only a small percentage of energy is transferred from one trophic level to the next. The rest is used to make ATP in the process of respiration and is eventually lost as heat.

▲ **Figure 36** Humming birds have a larger surface area to volume ratio than other larger mammals or birds and therefore lose a lot more heat relative to their size.

Note however that the values on Figure 34 are only generalisations. There are many factors that influence exactly how much energy is transferred at each stage.

- Mammals are **endothermic**. This means that they are able to keep their body temperatures more or less constant at a value between approximately 35 °C and 40 °C, depending on the species. This high temperature is a result of heat produced during metabolism. Crocodiles (Figure 35) are found in many parts of the tropics. They rely on their environment to maintain a high body temperature. More of the food they eat can therefore be converted into new cells and tissues, and less chemical potential energy will be lost in maintaining body temperature.
- The surface area to volume ratio of a small mammal or bird, such as the humming bird in Figure 36, is much larger than that of a large mammal or bird. Small mammals and birds therefore lose a lot more heat relative to their size and cannot convert as much of the food that they eat into new cells and tissues.
- In general, carnivores such as the mantid in Figure 37, convert the food they eat into new tissue more efficiently than do herbivores. Herbivores feed on plant material and plants contain a lot of substances, such as cellulose and lignin, that are difficult to digest. A much higher proportion of the food that a herbivore eats passes through the gut and is lost as faeces.

Rearing domestic livestock

Rearing livestock is a commercial business which needs to be profitable. Clearly, if a farmer is to run a successful business, he needs the maximum yields of milk from his milking herd, eggs from his hens, or meat from his

▲ **Figure 37** A mantid, a carnivorous insect.

livestock. In biological terms, he wants maximum net productivity. Look at the formula for net productivity:

$$\text{net productivity} = \text{gross productivity} - \text{respiration}$$

At its simplest, maximum net productivity involves manipulating conditions so that gross productivity is as high as possible and the loss due to respiration is as low as possible. In this next section, we will look at some of the factors associated with ensuring maximum net productivity of chicken.

Productivity and poultry farming

Commercially, chickens are either reared to produce meat or to lay eggs. Chickens reared to produce meat are known as broilers. Figure 38 shows growth curves for male and female broilers.

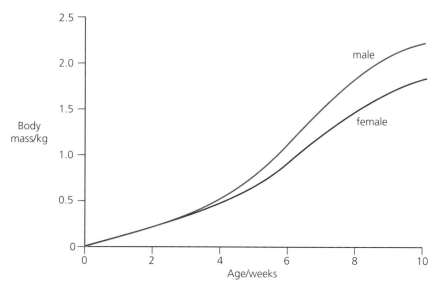

▲ **Figure 38** Growth curves for male and female broilers.

You can see from this graph that intensively reared broilers, fed with high energy, high protein food grow very rapidly. A modern broiler may be ready for marketing 7 to 8 weeks after hatching. At this age it is still growing and may not have even reached its maximum growth rate. The reason for slaughtering birds at this age can be seen if you look at the data in Table 4 opposite. Look at the figures for male birds. You can see from the last column that the mean mass of food eaten per kilogram gain in mass rises steadily. In other words, efficiency of food conversion is falling. This is mainly because the bird produces less protein-rich muscle and more body fat as it gets older.

Table 4 also shows that, by the time they have reached 10 weeks, female broilers have a smaller mean body mass and their efficiency of food conversion is lower than that of males. The difference in efficiency of food conversion may not seem very much – 2.5 kg of food per kilogram gain in body mass compared with 2.4 kg in males, but differences like this have a considerable influence on overall profit.

Q13

Trout reared in fish farms probably convert food into new tissue more efficiently than any other animal. Suggest why trout are able to convert so much of the food they eat into new tissue.

Q14

Intensively reared broilers are kept indoors under temperature-controlled conditions. Suggest how controlling temperature may increase productivity.

Sex	Age/weeks	Mean body mass/kg	Mean cumulative mass of food eaten/kg	Mean mass of food eaten per kilogram gain in mass/kg
Male	2	0.2	0.3	1.5
	6	1.0	2.0	2.0
	10	2.2	5.4	2.4
Female	2	0.2	0.3	1.5
	6	0.9	1.8	2.0
	10	1.8	4.5	2.5

Table 4 Growth and food consumption in broiler chickens.

Some ethical issues

Poultry scientists have discovered a gene that reduces the number of feathers in chickens. Broilers with this gene have fewer feathers and their necks are often bare. Because of this, they are called naked-neck broilers. Other scientists investigated the performance of naked-neck and normally feathered broilers in regions of Turkey that have different climates. Here is a summary of the abstract from the scientific paper in which the scientists reported their research.

Intensively reared chickens suffer at high environmental temperatures because their feathers prevent heat loss. Naked-neck broilers and normally feathered broilers were compared in three regions of Turkey where the climate differed. The mean temperatures in these regions were 19 °C, 28 °C and 32 °C.

The naked-neck broilers were found to have up to 20% fewer feathers than the normally feathered birds. They had the greatest mass of breast meat in all cases, between 2.5% and 10.9% higher than that of the normally feathered birds. The difference between the two groups of birds in the total body mass was also related to temperature. At the site with the hottest climate, the naked-neck broilers had a significantly greater body mass than the normally feathered group. At the site with the lowest temperature, body mass at 7 weeks was similar for birds in both groups.

In the hottest climates feed efficiency was about 9% higher in the naked-neck broilers. In these birds core body temperature was also lower. It is concluded that naked-neck broilers should be preferred in hot climates.

We will look at the way in which this investigation was carried out and consider some of the ethical issues involved.

■ Explain the link between the number of feathers and the productivity of broilers in hot conditions.

The abstract refers to the fact that naked-neck broilers do not suffer as much as normally feathered broilers from heat stress, and their body temperatures are lower than those of normally feathered birds. A lower body temperature means that rate of respiration is lower, so more of the chemical potential energy in the food goes towards producing new cells and tissues.

■ How might the results of this investigation benefit poultry farmers?

The information in the abstract suggests that farmers who raise broilers in hot climates could benefit from selecting naked-neck broilers. The broilers will be larger and will have a higher rate of growth. The efficiency of food conversion is also considerably higher.

Agricultural research such as that described in this abstract is funded both by public funding and by companies that obtain their income from poultry farming. Sometimes this raises ethical issues. In addition, individual scientists might have concerns about research such as this.

■ The mass of the breast meat of the birds in this investigation was expressed as a percentage of the total body mass after 10 hours without food. Explain the advantage of giving these measurements after 10 hours without food.

This enables a more reliable comparison because the birds might otherwise have different amounts of food in their guts. Clearly a bird that had just eaten would have a higher total body mass because of the food in its gut. This would result in the percentage of breast meat being lower than if the same bird had been starved first.

■ Is it ethically acceptable to starve the broilers in order to ensure that a more reliable comparison can be made?

There is no right and wrong answer to this question – much depends on your personal standpoint. Broilers are normally starved before slaughter. Part of the reason for this is that it lowers the probability of contamination of meat by bacteria from gut contents. It is also important, however, to consider the distress that 10 hours of starvation may cause.

If you return to the abstract you will see that it says that at the site with the hottest climate, the naked-neck broilers had a significantly greater body mass than normally feathered broilers. Let us look at the implications of the word 'significantly'.

■ How did the scientists who conducted this research make sure that they could claim that the difference in body mass between the two groups of broilers was 'significant'?

They would have had to analyse their results statistically. This would have involved comparing the results from suitably large samples with an appropriate statistical test.

■ Is it ethically acceptable to kill large numbers of birds in order to carry out a statistical test and determine whether differences are significant?

Again, there is no right or wrong answer to this question. For most people, a large-scale experiment that involves slaughtering huge numbers of birds would raise concerns. On the other hand, the findings of this investigation could provide cheaper food for many. There is also the question of what happens to the meat after slaughter. If it is used for human consumption, then the fate of these birds is no different from the fate of other broilers. If the meat is destroyed, that is another matter.

Pests and agricultural crops

Diseases such as those caused by viruses, bacteria and fungi, pests such as insects and slugs and snails, and weeds reduce the yield of agricultural crops. Table 5 shows estimated losses from diseases, pests and weeds in some vegetable crops.

In Table 5, potential yield refers to the yield that would have been harvested without any loss from diseases, pests or weeds. The data in the table show that the total loss for all of these crops is quite substantial, ranging from 13% in cauliflower to 34% in peas. They also show that different crops are affected in different ways. Diseases, for example, are very important in reducing potato yield, while the most important factor with peas is the presence of weeds.

The conclusions that may be drawn from the data in Table 5 are, however, limited. When we look at data such as these, we should bear the following in mind.

Crop	Loss as percentage of potential yield due to:			Total percentage loss
	diseases	pests	weeds	
Cabbage	4	9	5	18
Cauliflower	3	5	5	13
Onion	13	10	8	31
Pea	7	4	23	34
Potato	22	7	4	33
Tomato	12	7	5	24

Table 5 Estimated global losses from diseases, pests and weeds in selected vegetable crops.

Q15

It would not be helpful to calculate the mean percentage losses due to diseases, pests and weeds for all the crops in Table 5. Suggest **two** reasons why.

- The data are, at best, only crude estimates. It would not be possible for agricultural scientists to obtain accurate figures for crop losses on a global scale. Because of this, although it might be reasonable to conclude that the total percentage loss of potatoes is greater than that of cabbages, we would not be justified in stating that the percentage loss of potatoes was greater than that of, say, onions.
- The figures in the table are mean values. They give no idea of the fluctuations that occur from year to year or from region to region. For example, the mean value for loss due to insect damage in cauliflowers in the UK is about 3%. Local losses, however, may be much greater than this and values as high as 89% have been recorded.
- The figures in the table relate to yield. They give no indication of quality.

Controlling pests with chemical pesticides

Pests and weeds may be controlled with chemical pesticides. Pesticides may be divided into **insecticides** used to combat insect pests, **fungicides** which target the fungi that cause many plant diseases and **herbicides** which are used to kill weeds. Chemical pesticides work in different ways.

- **Contact pesticides** are sprayed directly on a crop. A contact insecticide is usually absorbed by an insect through its tiny gas-exchange pores called spiracles which are positioned along its body. Contact fungicides and herbicides are absorbed directly through the surface. Contact pesticides are relatively cheap, but they have to be reapplied. This is because their effect is short-lived. In addition, there are always some pests that avoid the pesticide.
- **Systemic pesticides** are also sprayed on the crop. They are absorbed by the leaves and transported around the plant. An insect, such as an aphid feeding on the sprayed plant, takes in the pesticide and is poisoned. Systemic insecticides are very effective because it is not necessary for the spray to come into contact with every insect. These pesticides are also selective in their actions. Systemic herbicides are absorbed by leaves and, because they are transported through the weed, kill all of its tissues, including underground parts such as the roots.
- **Residual pesticides** are sprayed on the soil or used to treat seeds before the crop is planted. They remain active in the soil and kill fungal spores, insect eggs and larvae and the seedlings of weeds when they germinate.

A good insecticide has three properties:

- it must be toxic or it will not kill its target organism
- it must be persistent so that its effects last
- it must not harm other organisms.

In the early 1950s, an insecticide called DDT was thought to have all these properties and its use became widespread. Coinciding with this, however, scientists noticed a sharp decline in the numbers of some species of bird. One of these was the peregrine falcon (Figure 39), a bird that feeds mainly on other birds.

Scientists investigated the effect of DDT on peregrine falcon eggshell thickness and strength in two samples of eggs.

▲ **Figure 39** Peregrine falcon. Research has shown that DDT affects breeding success in birds of prey.

- They obtained the first sample from museum collections. These were eggs laid between 1850 and 1942, before the use of DDT.
- They collected the second sample between 1970 and 1974. These eggs were from birds that had been exposed to DDT.

The graph in Figure 40 shows some of the results of the investigation.

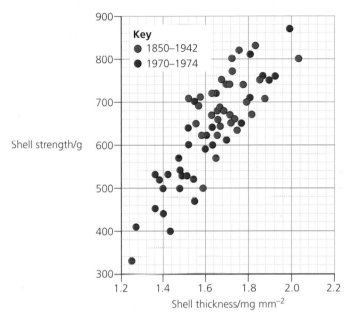

▲ **Figure 40** Shell thickness and strength in eggs from peregrine falcons laid between 1850 and 1942 and between 1970 and 1974.

Egg strength and DDT

1 a) The scientists measured shell strength by finding the mass required to push a needle through the shell, under standard conditions. Suggest one condition which would need to be standardised in making these measurements. Give a reason for your choice.

b) The museum eggs had had their contents removed so that only the shells remained. Explain the advantage of measuring their thickness in mg mm^{-2}.

2 What statistical test could you use to show:

a) whether there was a significant correlation between shell strength and shell thickness

b) whether there was a significant difference between the shell strengths of the two groups of eggs?

3 The differences between the shell strengths and the shell thicknesses of the two groups of eggs were both significant at the 0.05 probability level. Explain the meaning of 'significant at the 0.05 probability level'.

4 It was suggested that there was a link between shell thickness and hatching success of the eggs. Describe and explain this link.

5 Is it valid to conclude from the data in this investigation that DDT causes the shells of peregrine falcon eggs to be thinner? Explain your answer.

6 DDT does not break down readily in the body of a bird. It is stored in the bird's body fat. Explain why peregrine falcons were particularly susceptible to poisoning by DDT.

Controlling pests with biological agents

Biological control involves making use of natural enemies for pest control. These may be predators, parasites or pathogens. The object is to reduce the number of pests present and hence the damage they cause.

The first example of biological control involved a pest of citrus crops, the cottony cushion scale. This insect is native to Australia where it feeds on various wild plants as well as on introduced citrus trees. In Australia it is rarely a pest because it is attacked by various wasps, beetles and flies. Towards the end of the nineteenth century, citrus trees were introduced into California from Australia. Unfortunately there were cottony cushion scales present on some of these young trees. Without their natural enemies, the scale insects multiplied rapidly and soon became a major threat to the Californian citrus crops.

The US Department of Agriculture responded to this threat by sending one of its scientists to Australia. He found a variety of insects that fed on the scale insect and introduced one of these, a ladybird beetle, to California. The 500 or so beetles that were released spread and multiplied rapidly, effectively controlling the cottony cushion scale. Both species are still found in California but at very low population densities.

Before a pest can be successfully controlled by another organism, extensive research must be carried out. Figure 41 shows some of the main steps involved in setting up a biological control programme.

Q16

Serious outbreaks of cottony cushion scale now only occur in California where insecticides are used against other pests of citrus crops. Suggest why.

> Many animals and plants have been introduced to areas by humans and are not native to the area in which they have become pests. In their new environment they multiply very rapidly because they have few predators or parasites. Biologists search the area where the pest originated to find suitable organisms to use in a **biological control programme**.

> Trials are carried out to make sure that the control organism:
> • attacks the pest only
> • does not carry diseases that might spread to other native animals or plants
> • can establish itself and maintain its population in its new environment
> • will not itself become a pest due to a lack of predators or parasites.
> Great care is taken to ensure that the control organism does not escape before the trials are complete.

> The control organisms are bred or cultured in large numbers. They are then distributed and released.

> Agricultural scientists collect the necessary evidence to evaluate the programme and determine whether it has been successful.

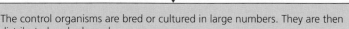

▲ **Figure 41** The main steps involved in a biological control programme.

The water hyacinth (Figure 42) is a plant which floats on the surface of the water of lakes and slow-flowing rivers. Parts of the plant break off very easily and grow into new daughter plants. It has been introduced into many parts of the world. Where no native animals feed on it, it spreads rapidly and is an important weed, choking waterways and interfering with transport and drainage.

Scientists discovered a beetle that feeds on water hyacinths. In Zimbabwe, water hyacinths are serious pests. Scientists investigated whether the beetle could be used to control water hyacinths. They grew the plants individually in pots in a glasshouse and added two pairs of beetles to each potted plant. They set up a second group of plants as a control. The graph in Figure 43 shows the mean leaf areas of the experimental plants and the control plants at monthly intervals.

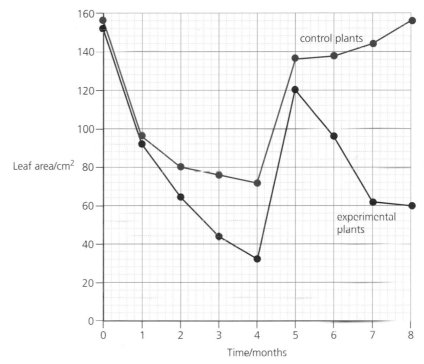

▲ **Figure 43** The effect of a species of beetle on mean leaf surface area of water hyacinth plants.

▲ **Figure 42** A water hyacinth in flower. Because of its attractive flowers, the water hyacinth has been introduced into many parts of the world. With no natural enemies, it rapidly becomes an important weed.

The scientists made two suggestions to explain the results obtained between months 4 and 5.

A The beetles that were present on the water hyacinth plants were not feeding at this time.
B The temperature in the glasshouse at this time caused the water hyacinth plants to grow very rapidly.

■ Use information from Figure 43 to explain which of these suggestions you think is more likely.

We will start by looking at suggestion **A**. If the beetles stopped feeding, it is likely that the surface area of the leaves of the plants on which they were found would increase. This would not explain, however, the increase in the surface area of the plants in the control pots where there were no beetles. In view of this, suggestion **B** seems more likely.

Table 6 shows some more results from this investigation.

Month	Mean number of leaves		Mean length of longest root/cm		Mean number of daughter plants	
	experimental plants	control plants	experimental plants	control plants	experimental plants	control plants
1	6.5	7.1	35	26	0	0
2	6.5	6.5	10	10	6	6
3	6.5	6.5	9	12	4	6
4	7.0	6.0	7	10	5	6
5	7.0	7.0	20	19	6	6
6	7.0	7.0	21	20	4	7
7	6.0	7.0	11	20	6	7
8	6.0	6.0	11	23	7	7

Table 6 Some effects of a species of beetle on the growth of water hyacinth plants.

■ Is this species of beetle likely to be of use in controlling water hyacinths in Zimbabwe? Use the results of this investigation to support your answer.

The answer to this question is not 'yes' or 'no'. It is a 'maybe'. If you compare the experimental plants with the control plants, you will see that after 8 months there is no difference in either the mean number of leaves per plant or in the mean number of daughter plants produced. This suggests that the beetle will not be an effective biological control agent. However, the evidence from the graph and the table shows that the experimental plants were smaller. They had a smaller mean leaf surface area and the mean length of the longest root was less. This suggests that the beetle might be of some use in controlling water hyacinth plants. We should also bear in mind that this investigation was carried out in a glasshouse. Conditions might be different in the field.

■ Suggest why this investigation was carried out in a glasshouse rather than in the field.

This was an important measure to prevent the beetles from escaping before all the trials were complete.

Integrated pest management

Integrated pest management (IPM) is a series of steps which involve evaluating specific situations and making decisions. IPM makes use of all available methods of control so that pests are managed in the most economical way and with the lowest risk to people and to the environment.

There are four main steps involved in integrated pest management.

1 **Monitoring and identifying pests.** Many of the insects found on crops do little harm. Some, such as ladybirds and ground beetles, may be beneficial because they are predators of important pests. In a similar way,

▲ **Figure 44** Potato blight on a potato leaf. This fungus can grow rapidly, destroying the whole plant.

not all weeds have a significant effect on crop yield. IPM programmes involve monitoring crops and identifying pests. Accurate identification means that pesticides are less likely to be used when they are not needed.

2 **Setting action thresholds.** Finding a single pest does not necessarily mean that control is necessary. Before taking further action, a threshold is set. This is the point at which a pest population or environmental conditions suggest that control will be necessary. Potato blight (Figure 44) is a fungus that attacks potatoes. It spreads rapidly in warm, humid conditions. By identifying these conditions, growers can target spraying to times when it will be most effective.

3 **Preventing pests reaching action thresholds.** The first step in pest control is to prevent it becoming established in a crop. Cultural methods, such as crop rotation and the planting of pest-resistant varieties, are often very effective in doing this. Not only are they cost-effective, but they also offer little risk to the environment.

4 **Controlling pests.** When monitoring shows that a pest population has reached the **action threshold**, a control method must be selected. Selection of an appropriate method of control involves considering both its effectiveness and the environmental risk. The least risky control methods are chosen first. These include methods such as weeding, biological control and the use of highly specific chemical pesticides. Only if these fail to work are other methods, such as spraying with non-selective pesticides, used.

Chapter 3
Nutrient recycling

Scientific discoveries often benefit human communities. Vaccines and antibiotics, for example, have had a huge impact on the quality of human life. Most scientific research doesn't lead to major discoveries such as these. The small-scale benefits it brings, however, can still be very important.

One of the problems that almost all human communities face is how to get rid of the enormous quantities of waste they produce. Some of this waste can be composted. Composting uses microorganisms to break down organic material. The compost that is produced can be added to the soil as a conditioner and fertiliser. Composting has been carried out for hundreds of years – the ancient Romans left written records of composting – but scientific research has led to discoveries that mean that we can now make compost more efficiently and on a much larger scale (Figure 1).

Figure 2 shows how compost is made. The main steps are summarised below.

▲ **Figure 1** Composting on a commercial scale.

- Suitable organic material, for example grass clippings, vegetable waste and even coffee grounds and cardboard, is collected into a heap. In small-scale garden compost heaps soil invertebrates break this material down into smaller particles. In large-scale composting, it is broken down by mechanical grinders and choppers.

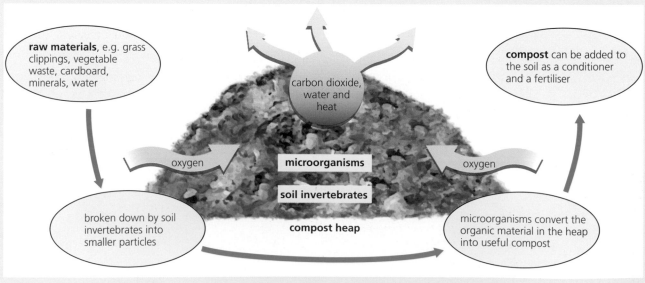

raw materials, e.g. grass clippings, vegetable waste, cardboard, minerals, water

carbon dioxide, water and heat

compost can be added to the soil as a conditioner and a fertiliser

oxygen

microorganisms

oxygen

soil invertebrates

broken down by soil invertebrates into smaller particles

compost heap

microorganisms convert the organic material in the heap into useful compost

▲ **Figure 2** How compost is made.

- Soil microorganisms colonise this heap. They are called **mesophils** because they live at temperatures of between 10 °C and 45 °C. As they multiply and respire, the compost heap heats up.

- The mesophils are gradually replaced by **thermophils**. Thermophils are microorganisms that live in conditions where the temperature is high – in this case, between approximately 50 °C and 65 °C. This is the **active phase** of composting and these thermophilic microorganisms are mainly responsible for converting organic material in the heap into useful compost.

- The last stage is the **curing phase**. The temperature falls and mesophils again colonise the heap. During the curing phase the compost matures until it finally becomes suitable for adding to the soil.

Compost feedstock	Typical C:N ratio
Autumn leaves	60:1
Cardboard	500:1
Coffee grounds	20:1
Maize stalks	65:1
Pig manure	25:1
Vegetable waste	15:1

Table 1 Typical C:N ratios for some common compost feedstocks.

In this chapter, we shall look at the recycling of chemical elements in ecosystems. We shall follow the passage of carbon and nitrogen through different trophic levels and consider the role of microorganisms in converting organic substances into the inorganic substances and ions that are taken up by plants.

The science of composting

The ratio of carbon to nitrogen in compost is known as the **C:N ratio**. It is a very important aspect of compost feedstock – the organic material that is originally gathered into the compost heap. Scientists have found that the ideal starting point is a C:N ratio of about 30:1.

- If the C:N ratio is greater than this, the soil microorganisms in the heap use most of the nitrogen for their own metabolic needs. There will only be a low concentration of useful nitrogen-containing substances in the compost.
- If the C:N ratio is lower than this, there will be surplus nitrogen and most of this will be lost to the atmosphere as ammonia.

Table 1 shows the C:N ratios of some common compost feedstocks.

Information such as that in Table 1 was originally obtained from experimental work carried out by ecologists investigating decomposition in natural ecosystems. By making use of this work, compost is now produced efficiently on an industrial scale.

In the last chapter we saw how energy was transferred from one trophic level in a food web to the next. Energy transfer is inefficient and all the energy entering an ecosystem is eventually lost as heat. Energy is described as a renewable resource. Elements such as carbon and nitrogen are described as non-renewable. There is only a certain amount of them and they are continuously recycled.

Q1

The main substance present in maize stalks is cellulose. Explain why maize stalks have a high C:N ratio.

Q2

When earthworms feed on a compost feedstock, they decrease its C:N ratio. Suggest how earthworm activity leads to a decrease in the C:N ratio.

Dead leaves and decomposition

We saw in the last section that in small-scale garden compost heaps soil invertebrates break plant material down into smaller particles. This also happens in natural habitats, such as woodlands. These organisms include arthropods, such as insects and woodlice, nematode worms and earthworms. The graph in Figure 3 compares the biomass of these organisms and of fungi and bacteria in a British woodland at different times of the year.

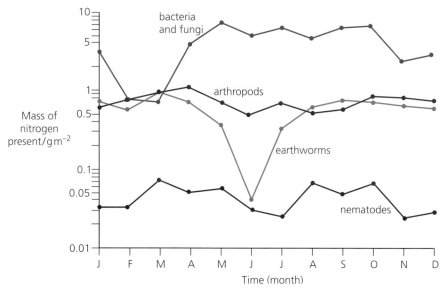

▲ **Figure 3** Variation in the biomass of different groups of organisms involved in breaking down dead leaves at different times of the year.

This graph is quite complex. Before we look at the conclusions that we can draw from it, we need to make sure that we understand how the data have been presented.

- **The biomass has been given as the mass of nitrogen present. Explain why nitrogen content can be used as a measure of biomass.**

  Nitrogen is present in many of the substances that make up the structure of soil organisms. These substances include protein, DNA and chitin, which forms the hard outer layer of insects. The proportion of nitrogen-containing substances must obviously be more or less constant in these organisms if it is to be an accurate and reliable measure of body mass.

- **The data in the graph have been plotted on a log scale. Suggest the advantage of using a log scale.**

  A log scale lets a greater range of data be plotted than an arithmetic scale. If you look carefully at the y-axis, you will see that it runs from 0.01 to 10 g m^{-2}. It also shows that the mass of nitrogen present in nematodes fluctuates from approximately 0.02 to 0.08 g m^{-2}. If the data on nematodes had been plotted together with all the other information on a scale with an arithmetic axis, they would have appeared as a straight line. You wouldn't have been able to see the fluctuations.

■ Suggest an explanation for the change in biomass of bacteria and fungi between February and June.

Leaves fall and accumulate in the autumn so there will be plenty of dead leaves available. The most likely explanation for the increase in biomass is that the soil is getting warmer and the enzyme-controlled processes associated with leaf breakdown and the growth of microorganisms are faster in these conditions.

Ecologists investigated the importance of arthropods and microorganisms in the decomposition of leaves. Different numbers of woodlice were added to containers, each with the same mass of dead oak leaves. The ecologists measured the rates of respiration of the microorganisms in these containers and in a control container. The results are shown in Figure 4.

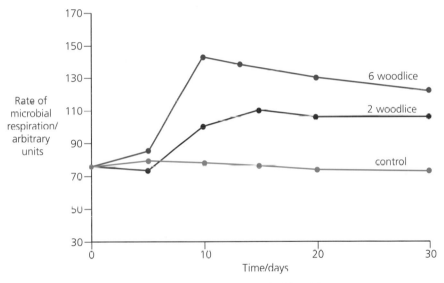

▲ **Figure 4** The effect of woodlouse activity on the microbial breakdown of leaves.

■ Describe how the control container should have been set up in this investigation.

It should be set up exactly the same as the other containers and exposed to the same environmental conditions. The only difference should have been that it did not contain any woodlice.

■ What does Figure 4 show about the effect of woodlouse activity on the microbial breakdown of leaves?

To answer this question you need to compare the containers in which woodlice were added to the control container. You can see that microbial respiration in the control chamber remains more or less constant. After 5 days, the rate of microbial respiration in both of the containers with woodlice increases and is higher than in the control. In addition, there is a greater rate of microbial respiration when six woodlice have been added than when there are only two woodlice in the container.

■ Suggest how the presence of woodlice could have caused the increase in the microbial respiration rate that occurred between 5 and 10 days.

There are two possible answers to this. The woodlice break up the leaves so there is more surface area to be colonised by microorganisms. As we saw when we were considering the formation of compost at the start of this chapter, feeding activity by soil invertebrates such as woodlice also decreases the C:N ratio of the leaf litter. This makes the environment more suitable for bacteria, hence their numbers increase and so does the rate of microbial respiration.

Nutrient cycles

Living organisms, such as animals and plants, require many different chemical elements. Plants take these elements up either as carbon dioxide in the case of carbon, or as ions from the soil. Inside the plant they are involved in various chemical reactions and are eventually converted into organic substances that form plant cells and tissues. Consumers obtain their supplies of these elements from plants or from other animals that feed on plants. A snail, for example, digests the complex organic molecules in its plant food and absorbs the products through its gut wall. In this way, elements are passed from organism to organism along the food chains that make up a food web (Figure 5).

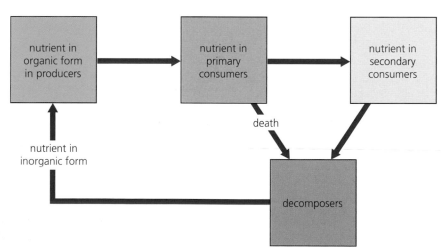

▲ **Figure 5** A simple nutrient cycle. All elements – carbon, nitrogen, phosphorus or iron – are cycled in this basic way. The detail of the process differs in different nutrient cycles.

Those organisms, or parts of organisms, that are not eaten as food eventually die and decompose. Microorganisms, such as fungi and bacteria, break down the complex organic molecules that form the dead material (Figure 6). They absorb some of the products but the rest are released as inorganic substances that can be taken up by plants.

This is the basic nutrient cycle. Elements are taken up by plants and passed from organism to organism in the various trophic levels. Death and decomposition result in microorganisms making these elements available to plants again.

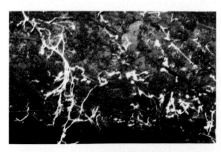

▲ **Figure 6** These white thread-like structures are fungal hyphae covering the surface of a piece of rotting wood. The fungi are decomposers. Their hyphae secrete enzymes which break down organic substances such as lignin in wood.

The phosphorus cycle

1 Nucleic acids contain phosphorus. Name one other phosphorus-containing substance found in producers.

2 Explain how phosphorus in substances found in producers enters the body of a primary consumer.

3 Describe the part played by microorganisms in making the phosphorus in the body of a secondary consumer available to a plant.

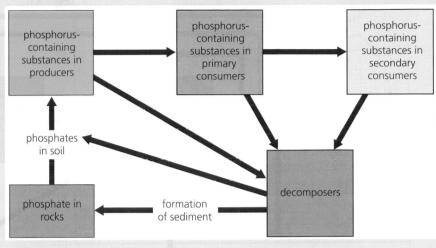

▲ **Figure 7** A simple phosphorus cycle.

4 Marine fishing transfers about 50 million tonnes of phosphorus to the land each year. Explain how.

5 Other than from marine fishing, suggest two other ways in which the concentration of phosphates in the soil and in lakes and rivers is increased by human activity.

The carbon cycle

Plants take up carbon dioxide from the atmosphere. It diffuses into photosynthesising cells where it is converted to sugars and other organic substances. Primary consumers feed on plants, and carbon-containing substances in their food are digested. Smaller, soluble molecules are produced and these are absorbed. They are built up into the carbohydrates and other large molecules that make up the tissues of the primary consumers. The same thing happens when a primary consumer is eaten by a secondary consumer. In this way, just as in all nutrient cycles, the element is passed from one trophic level to the next through the food web.

The next step involves the decomposers. Many soil bacteria and fungi are **saprobionts**. They secrete enzymes that break down the complex organic substances which make up dead animals and plants. The smaller molecules are absorbed by the microorganisms and some of them are used in respiration. Respiration releases carbon dioxide which can be used by plants in photosynthesis. The cycle (summarised in Figure 8, overleaf) is basically complete.

There is one additional factor that we need to consider, however. We do not have to wait for an animal or plant to die before any of the carbon stored in its tissues is released. All organisms respire and release carbon dioxide into the atmosphere.

■ Explain why the values for the rate of carbon dioxide absorption at the soil surface are always negative.

The soil surface contains the lower stems and some of the roots of the wheat plants. It also contains many soil microorganisms. Only respiration is taking place here, so carbon dioxide is being produced, not taken up.

Q3

Explain why, on a warm day in spring, you would expect the rate of photosynthesis of the plants in a community to be greater than the rate of respiration of all the organisms in the community.

Investigations such as this help to show how the concentration of carbon dioxide varies over a 24-hour period. During daylight, plants will be photosynthesising. All the organisms in the community will be respiring. The rate of photosynthesis is in general greater than the rate of respiration so carbon dioxide concentration falls as the gas is absorbed by the plants. At night, only respiration is occurring. Respiration produces carbon dioxide as a waste product so the concentration of carbon dioxide rises.

Carbon dioxide, methane and climate change

Energy from the Sun warms the Earth's surface. Some of this energy is reflected back, but it is at a longer wavelength than the incoming energy. Carbon dioxide, water vapour and other gases prevent much of this long-wavelength energy escaping from the atmosphere. They reflect it back to the Earth's surface. This is called the **greenhouse effect** because it involves exactly the same principle as that which results in the temperature inside a greenhouse being higher than that outside.

The greenhouse effect is not something that is caused by humans. It is a consequence of the Earth having an atmosphere containing gases, such as carbon dioxide, methane and water vapour. The greenhouse effect is one of the reasons why life has been able to evolve on Earth because it helps to keep temperatures more stable than they would otherwise be. Human activities have enhanced, not produced, the greenhouse effect by leading to an increase in the atmospheric concentration of carbon dioxide, methane, oxides of nitrogen and other gases. All these result in less long-wavelength energy being radiated into space.

One of the important features of the Industrial Revolution was a switch from using sustainable fuels, such as wood and charcoal, to the use of coal and, later, oil. From the mid-nineteenth century, burning of these fossil fuels and the clearing of forests have increased the concentration of carbon dioxide in the atmosphere from approximately 280 ppm to nearly 400 ppm. Figure 10 shows some evidence for this increase. It is based on analysis of gas samples trapped in Antarctic ice.

The concentrations of other greenhouse gases have also risen over this period. One of these is methane (CH_4). Part of the explanation for the increase in methane is the increase in the intensive cultivation of rice. Rice grows in swampy paddy fields and the bacteria that live in these anaerobic conditions produce large amounts of methane. Another part of the explanation is the increase in the number of cattle. Cattle are ruminants and rely on their gut bacteria to break down cellulose in the plant material on which they feed. A waste product of this process is methane. It is estimated that digestion in a cow produces approximately 40 dm³ of methane per day.

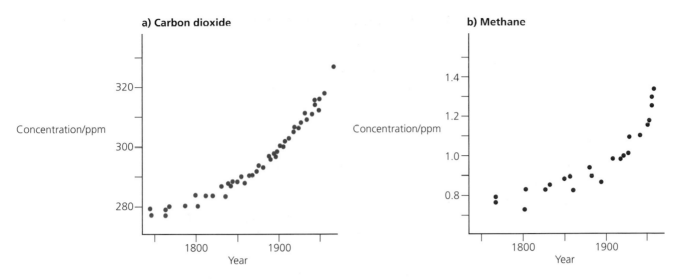

a) Carbon dioxide

b) Methane

▲ **Figure 10** Graphs showing the change in concentration of **a)** carbon dioxide and **b)** methane in gas samples trapped in the Antarctic ice.

Computer models have been used to predict the effect of an increase in greenhouse gases on global temperatures. Although their predictions vary, most models predict a rise of between 2.3 °C and 5.2 °C by the year 2100. Most of the variation is due to the way in which cloud cover is modelled. Water vapour is also an important greenhouse gas. If we ignore any changes in rainfall – these are very difficult to predict – an increase in mean surface temperature of just 3 °C would give London the climate of the Portuguese capital, Lisbon. At this temperature, we could grow grapes, citrus crops and olives easily!

Climate change and insect pests

You use physiological mechanisms such as shivering, sweating and the dilation of blood vessels to maintain a body temperature that is more or less constant. Winter or summer, your core body temperature is around 37 °C. An insect's core body temperature, however, varies much more and closely parallels that of its environment. A rise in environmental temperature, therefore, results in an increase in an insect's core body temperature. This, in turn, affects features such as activity, growth and reproductive rate. Look at the graph in Figure 11, overleaf.

The curve on this graph shows the effect of a constant environmental temperature on the hatching time of fruit fly eggs. This is the time from when the eggs are laid to when they hatch.

■ **Describe how environmental temperature affects the hatching time of fruit fly eggs.**

An increase in the temperature of the environment results in a decrease in hatching time. The fall is rapid at first but then slows down. Above a temperature of approximately 25 °C, environmental temperature has very little effect on hatching time.

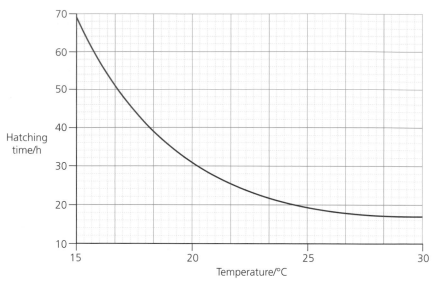

▲ **Figure 11** The effect of temperature on egg development in the fruit fly.

■ Suggest an explanation for the decrease in hatching time between 15 °C and 20 °C.

The development of a fruit fly larva inside an egg involves biochemical reactions. These are controlled by enzymes. An increase in temperature from 15 °C to 20 °C results in an increase in the rates of these reactions.

Scientists have made different predictions about how an increase in carbon dioxide concentration might change the mean global temperature. A widely held view is that doubling the mean carbon dioxide concentration is likely to produce a change in mean global temperature of between 2 °C and 4 °C.

■ How would you use the data in the graph to predict the likely effect of a rise in mean temperature of, say, 3 °C on hatching time in fruitflies?

Let us assume that the mean temperature is 16 °C. By reading the graph, you can see that the eggs take 57 hours to hatch. At a temperature 3 °C higher, that's 19 °C, the eggs will hatch in 35 hours. You can see that a relatively small increase brings about a large decrease in hatching time.

■ Explain why you should be careful in drawing general conclusions from this investigation about the effect of an increase in mean temperature on insect pests.

There are at least two important factors in the design of this investigation that suggest care must be taken in using the results to predict how an increase in mean global temperature might affect insect pests. The investigation involved a single species of insect, the fruitfly, and this species is not a crop pest. Different species will probably show different responses to temperature. Furthermore, this investigation was designed to show the effect of a constant environmental temperature on hatching time. Temperatures under field conditions are not constant. They will vary and this is bound to affect hatching time.

Other scientists have investigated the effect of temperature on the time that it takes insects to complete different stages of their life cycles. As a result of this work, it seems likely that an increase in mean temperature may affect insect pests in several ways.

- They will complete their life cycles in a shorter time. This means that there may be more generations in a year.
- A higher number of pests will survive over winter. They will emerge and infect crops earlier.
- Many tropical and sub-tropical pests will spread into areas where, at present, the climate is cooler.

The corn borer is an important pest of maize in many parts of the world. It is the larva of a moth that feeds inside the stem of a maize plant. The maps in Figure 12 show the distribution and number of generations in a year of corn borers in Western Europe. Figure 12 a) shows the information for the mean temperature between 1951 and 1980. Figure 12 b) shows the predicted effect of a 1 °C rise in mean annual temperature. If you compare these maps you will see that there is a northward spread in the potential distribution. In many areas, it is also likely that there will be an extra generation each year.

a) Mean temperature, 1951–1980

b) Predicted effect of an increase in mean temperature of 1 °C

Key
Generations per year

- one
- two
- three
- four

▲ **Figure 12** The effect of a predicted 1 °C rise in mean temperature on the distribution and number of generations per year of the corn borer, a pest of maize.

Viruses affect plants as well as animals. Plant viruses reduce crop yields. Aphids are small insects that feed by inserting their sharp mouthparts into the cells of the plant. Beet yellows is a virus that affects sugar beet. It is spread from plant to plant by feeding aphids. Scientists can predict the incidence of beet yellows virus from the number of days between the beginning of January and the end of March when the temperature falls below 0 °C and the mean weekly temperature in April (Figure 13, overleaf). Temperatures in January, February and March determine the number of aphids that survive the winter. The mean weekly temperature in April influences the rate at which the aphids on the crop multiply and spread to other crops.

Q4

Explain how completing its life cycle in a shorter time affects the damage a particular pest may do to a crop.

Q5

Explain how an increase in global mean temperature might increase sugar beet losses from beet yellows virus.

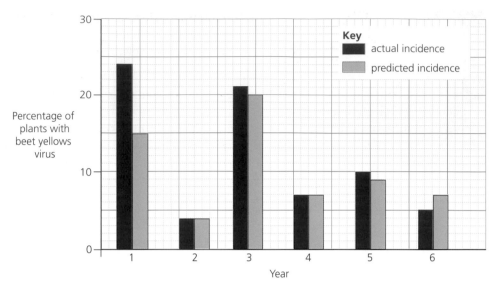

▲ **Figure 13** The incidence of beet yellows virus in sugar beet compared with the incidence predicted from temperatures in the preceding winter and spring.

Climate change and the distribution of wild animals and plants

In the introduction to Chapter 2, we looked at how a computer model helped biologists to predict the effects of an increase in carbon dioxide concentration and climate change on forests in North America. Butterflies have very specific habitat requirements and short life cycles. This means that they respond quickly to environmental changes, such as global warming and habitat loss.

Monitoring butterflies

Most of the species of butterfly found in Britain are too numerous to count individually. If we want to identify **trends** in the numbers and distribution of butterflies we need methods of monitoring populations that do not involve counting them all. We do this either by recording distribution or by measuring their abundance.

To record their **distribution** the person carrying out the survey decides when and where to record the species present. Records are then used to map distribution. The whole of Britain is covered by Ordinance Survey maps and can be divided into 10 km squares. We map distribution records on these squares and it provides us with information about the range of a species. We can then compare the present distribution with the distribution in the past and identify changes.

Look at Figure 14. It shows a 10 km square that has been divided into 100 smaller 1 km squares. The red spots show the presence of populations of a particular species of butterfly. If you count them you will see that there are ten 1 km squares that contain populations of this butterfly.

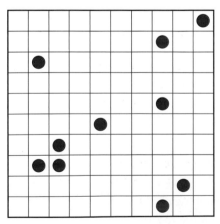

▲ **Figure 14** A 10 km square that has been divided into 100 smaller 1 km squares. The size of the squares on which we map distribution is important in identifying changes in populations.

Suppose that there is a decline in numbers of this species and it is now present in only five 1 km squares. A survey based on mapping distribution in 1 km squares will clearly show this decline. A survey based on mapping distributions in 10 km squares, however, will show no decline as the butterfly will still be present in the larger square. Larger squares, therefore, underestimate changes in butterfly populations (Figure 15).

When recording **abundance**, observers walk along fixed routes every week and count every butterfly within 5 metres. They walk this transect under standard conditions, such as at the same time of day and when the Sun is shining. Because the method is standardised, it is repeatable and the results can be analysed statistically. Techniques such as this have shown an increase in the numbers of the speckled wood butterfly, for example, as its range has spread further north (Figure 16).

▲ **Figure 15** A meadow brown butterfly. The distribution of this butterfly shows little change at the 10 km square level. The use of 1 km squares, however, has shown a marked decline in butterfly numbers in some parts of the country.

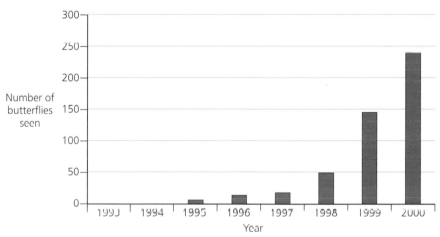

▲ **b)** A bar chart showing how the speckled wood butterfly has changed in abundance in a nature reserve in Cumbria.

▲ **Figure 16 a)** A speckled wood butterfly.

The nitrogen cycle

The general features of the **nitrogen cycle** differ very little from the basic nutrient cycle shown in Figure 5. Plants take up nitrate ions from the soil. The nitrates are absorbed into the roots by **active transport** and are used to produce proteins and other nitrogen-containing substances in plant cells. Primary consumers feed on plants and the proteins in their food are digested to form amino acids. These amino acids are absorbed from the gut and built up into the proteins that form the tissues of the primary consumers. The same thing happens when a secondary consumer eats a primary consumer. In this way, nitrogen is passed from one trophic level to the next through the food web.

When organisms die, the nitrogen-containing substances they contain are digested by **saprobiotic** bacteria. Nitrogen from consumers is also made available to saprobiotic bacteria through excretory products, such as urea. These saprobiotic bacteria release ammonia, so the process is called **ammonification**. Another group of bacteria, the **nitrifying bacteria**, then convert ammonia to nitrites and nitrates. The complete nitrogen cycle is summarised in Figure 17.

Q6

In what ways is the nitrogen cycle shown in Figure 17 similar to the carbon cycle?

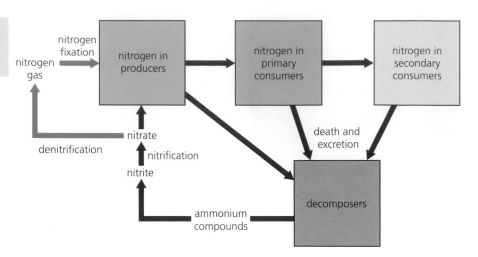

▲ **Figure 17** The nitrogen cycle.

Under anaerobic conditions, such as those that are found when soil becomes water-logged, **denitrifying bacteria** are found in large numbers. These bacteria are able to use nitrate in place of oxygen as an electron acceptor in the respiratory pathway. This reaction involves reduction of nitrate to nitrogen gas. This nitrogen escapes into the atmosphere so is no longer available to plants.

Nitrogen gas may be made available to plants again by **nitrogen fixation**. This term refers to a number of processes which involve the conversion of nitrogen gas to nitrogen-containing substances. The processes all require a considerable amount of energy. A nitrogen molecule consists of two atoms and the molecules have to be split into the atoms before nitrogen fixation can take place. Energy is required for this. Methods of fixing nitrogen include:

- **Lightning.** During a thunderstorm, the electrical energy in lightning allows nitrogen and oxygen to combine to form various oxides of nitrogen. Rain washes these substances into the soil from where they can be taken up as nitrates by plants. In the UK, the amount of nitrogen fixed by lightning is small. In some parts of the tropics, however, violent thunderstorms are common and this is an important way of fixing nitrogen.
- **Industrial processes.** The Haber process combines nitrogen and hydrogen to produce ammonia. The reactions involved take place at high temperatures and pressures in the presence of a catalyst. A lot of the ammonia produced by the Haber process is used to make fertilisers that are added to the soil.
- **Nitrogen-fixing microorganisms.** Many microorganisms are able to fix nitrogen. Some live free in the soil. Others are associated with the roots of leguminous plants, such as peas and beans, clover and lupins.

The biochemical reactions associated with nitrogen fixation are complex, but they can be summarised by the following equation.

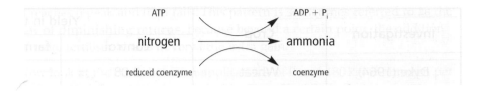

Nitrogen is reduced to ammonia. The reaction is catalysed by the enzyme nitrogenase. Nitrogenase, however, does not function in the presence of oxygen and many nitrogen-fixing organisms have adaptations which ensure that anaerobic conditions exist in the parts of cells involved in nitrogen fixation.

Fertilisers and productivity

When organisms living in natural ecosystems die, they decompose. Soil microorganisms convert, for example, the nitrogen in organic substances such as proteins and nucleic acids to nitrates, and these are taken up by the producers. There is a continual recycling of nutrients. In agricultural ecosystems, however, a large part of the biomass that is produced is harvested and removed. This is true whether we are considering crop plants, such as wheat or potatoes, or animals, such as dairy cattle and sheep. Unless the nutrients in this biomass are replaced, their concentration in the soil decreases and crop productivity falls.

There are two ways in which mineral ions such as those of nitrogen, phosphorus and potassium can be added to the soil. We can either use **inorganic fertilisers** or **organic fertilisers**, such as farmyard manure. Inorganic fertilisers:

- have a guaranteed composition making it easier to determine rates of application and predict the effect on crop yield
- are concentrated sources of nutrients and can therefore be applied in smaller amounts. This saves on transport costs and on the damage done by heavy machinery compacting the soil and crushing the crop
- are clean and convenient to handle and apply evenly.

Organic fertilisers, on the other hand:

- are mixtures of substances and may contain trace elements, substances that are important to plants in small amounts
- add organic matter to the crop. This may improve soil structure, reducing erosion and improving water-holding properties
- release the nutrients they contain over a longer period of time.

It is clear from various investigations that regular dressings with farmyard manure can benefit a crop and may produce increases in yield similar to those obtained with inorganic fertiliser. Table 2 overleaf shows the results obtained from a number of different investigations.

The first column in this table gives the name of the scientist or scientists responsible for carrying out the investigation and publishing the results. Comparing the work of different scientists allows us to see that similar findings have been reported by others. This helps to make sure that the conclusions that we draw are robust.

Q7

Explain why organic fertilisers release the nutrients they contain over a longer period of time than do inorganic fertilisers.

Q4

Which of the alleles in Table 2, I^A, I^B or I^O, is recessive?

Genotype	Polypeptide present in plasma membrane of all red blood cells	Name of blood group
I^AI^A	antigen A	group A
I^AI^O	antigen A	group A
I^BI^B	antigen B	group B
I^BI^O	antigen B	group B
I^AI^B	antigen A and antigen B	group AB
I^OI^O	neither antigen A nor antigen B	group O

Table 2 The possible genotypes for the human ABO blood group and the phenotypes associated with them. Here, blood group is the phenotype.

I^AI^A person, a heterozygous I^AI^O person has only antigen A in the plasma membranes of their red blood cells and is also blood group A. From this we conclude that the I^A allele is dominant and the I^O allele is recessive.

The third and fourth rows of the table tell a similar story about the I^B and I^O alleles. We conclude that the I^B allele is dominant and the I^O allele is recessive. Not surprisingly, only when the genotype is homozygous I^OI^O does the effect of the I^O allele show in the phenotype.

Q5

a) What is an allele?
b) Alleles I^A and I^B are co-dominant, but the I^O allele is recessive. Explain the meaning of this statement.

Now look at the fifth row of the table. In the heterozygous genotype I^AI^B, both antigen A and antigen B are produced in the plasma membranes of red blood cells. When both the alleles of a gene in a heterozygote show their effect in the phenotype, we call them **co-dominant** alleles.

Monohybrid inheritance involving only dominant and recessive alleles

In studying **monohybrid inheritance**, we follow the inheritance of a single character that is controlled by one gene. As we learn more about gene action, we find that few characteristics are actually controlled by only a single gene. Figure 9 shows examples of Fast Plants®, a form of *Brassica rapa* that was bred for research activities and is often used in school and college investigations because it has a very short life cycle. The plants on the right produce a purple pigment, called anthocyanin, in all parts of the plant. The plants on the left do not produce anthocyanin.

Constructing a genetic diagram to explain a monohybrid cross

Figure 10 shows what happens when a homozygous Fast Plant® that produces anthocyanin is crossed with a plant that does not produce anthocyanin. The dominant allele for anthocyanin production is given the symbol **A**; the recessive allele is given the symbol **a**. The **genetic diagram** has been laid out in a particular way.

The first row in the diagram shows the phenotypes of the parents; in this case one can produce anthocyanin and one cannot. The next line shows the genotypes of the parents.

▲ **Figure 9** Fast Plants® normally produce a purple pigment like those on the right of the photograph. These plants have a dominant allele that enables them to produce anthocyanin. The plants on the left of the photograph are homozygous for a recessive allele of this gene. They do not produce anthocyanin.

Constructing a genetic diagram to explain a monohybrid cross

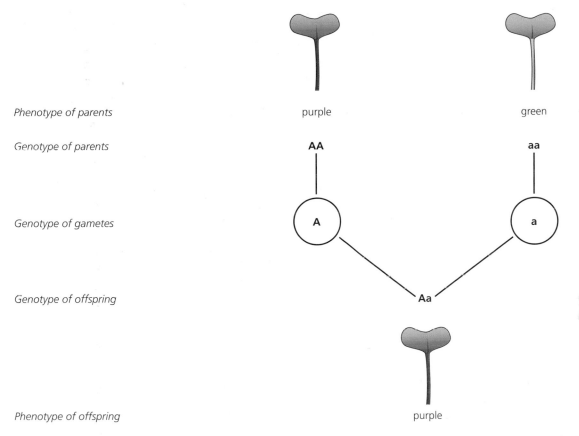

Phenotype of parents	purple	green
Genotype of parents	**AA**	**aa**
Genotype of gametes	**A**	**a**
Genotype of offspring	**Aa**	
Phenotype of offspring	purple	

▲ **Figure 10** A genetic diagram showing monohybrid inheritance of anthocyanin production when a purple Fast Plant® is crossed with a green Fast Plant®. You should use the layout of this genetic diagram when answering questions involving monohybrid inheritance.

■ **How do we know that the genotype of the parent that cannot produce anthocyanin is aa?**

We know that the parent that cannot produce anthocyanin must be homozygous, since the allele that results in no anthocyanin being produced is recessive (**a**). If the dominant allele (**A**) was present in the genotype, the plant would be able to produce anthocyanin. The next row of the genetic diagram shows the genotype of the gametes that each parent can produce. It helps to make the diagram clear if we put these genotypes in a circle.

■ **Why do the genotypes of the gametes contain only one allele, A or a?**

Gametes are always haploid because meiosis occurs when parents produce gametes. Therefore, the gametes produced during monohybrid inheritance will always contain only one of the alleles controlling the character we are investigating. The next row in Figure 10 shows the genotype of the offspring. Since all the gametes of one parent contain the **A** allele and all the gametes of the other parent contain the **a** allele, all the offspring must have the genotype **Aa** – they are heterozygotes. Their phenotype shows the effect of the dominant **A** allele, so they are all able to produce anthocyanin.

Figure 11 on the next page shows what happens if these heterozygotes are allowed to interbreed. It uses the same layout as Figure 10.

4 Genes, alleles and inheritance

■ Use your knowledge of meiosis to explain why you would expect each parent to produce equal numbers of gametes containing the **A** or **a** allele in Figure 11.

During prophase 1 of meiosis, homologous chromosomes pair together. They are then pulled apart, one to each end of the spindle, when spindle fibres contract during anaphase 1. This means that each gamete contains only one of the chromosomes from each homologous pair and half the gametes will have one of the chromosomes and half will have the other chromosome. One chromosome carries the **A** allele and the other carries the **a** allele, so the gametes with **A** and **a** will be produced in equal numbers.

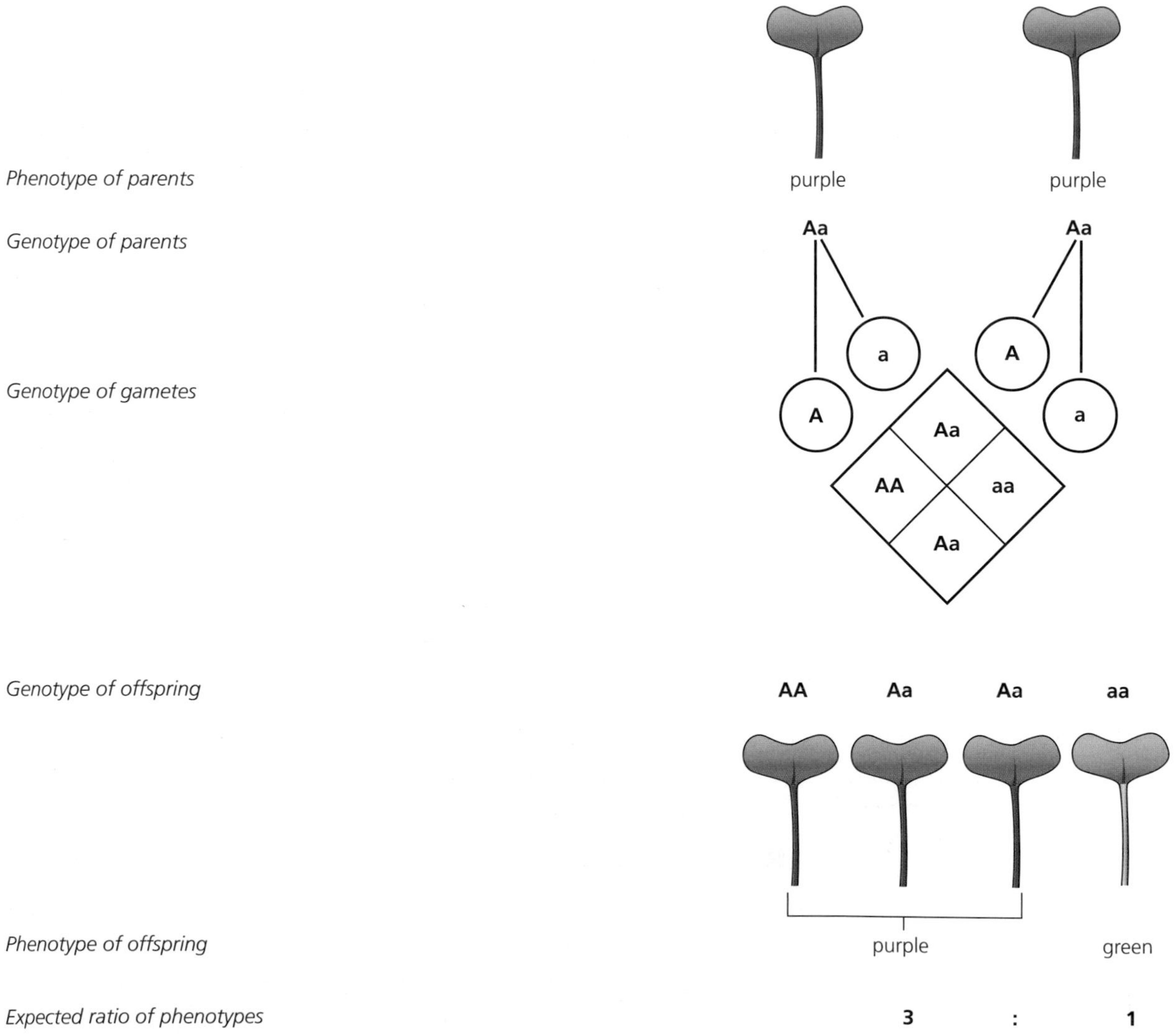

Phenotype of parents   purple   purple

Genotype of parents   **Aa**   **Aa**

Genotype of gametes   **a**   **A**   **A**   **a**
      Aa   **AA**   **aa**   **Aa**

Genotype of offspring   **AA**   **Aa**   **Aa**   **aa**

Phenotype of offspring   purple   green

Expected ratio of phenotypes   3   :   1

▲ **Figure 11** A genetic diagram showing the result of a cross between two heterozygous Fast Plants®. In a monohybrid cross involving a dominant and a recessive allele, a ratio of 3:1 is always expected in the offspring of heterozygous parents.

Constructing a genetic diagram to explain a monohybrid cross

Look at the fourth row in Figure 11. Instead of using the notation we used in Figure 10 to show fertilisation, we have used a Punnett square.

■ **Suggest why a Punnett square has been used in Figure 11.**

If we represented fertilisation by drawing lines between the two gametes from one parent and the two gametes from the other, we would end up with a network of crossing lines. Not only does this look untidy, it can confuse us as we complete the genetic diagram. We are much less likely to make a mistake, and so get our answer wrong, by using a Punnett square.

■ **How many different genotypes and phenotypes are present in the Punnett square in Figure 11?**

In a monohybrid cross between two heterozygous parents, there are three possible genotypes among their offspring. In this case, they are AA, Aa and aa. Since the A allele is dominant, plants with genotypes AA and Aa can produce anthocyanin, so there are only two phenotypes.

■ **The expected ratio of phenotypes in Figure 11 is shown as 3:1. What assumption has been made about fertilisation in order to make this prediction?**

We have assumed that fertilisation between the gametes of the parents occurs at random. If there are a large number of fertilisations, we expect gametes containing an A allele from one parent to fuse with equal frequency with gametes containing the A allele or a allele from the other parent. This gives us the expected ratio of phenotypes in the offspring.

Monohybrid inheritance in *Drosophila*

Figure 1 (page 99) shows an adult female fruit fly (*Drosophila melanogaster*). This fly has long wings. Other adults have small, stunted wings, called vestigial wings. Wing length is controlled by a single gene that has two alleles – one resulting in long wings and one resulting in vestigial wings.

Two heterozygous, long-winged *Drosophila* were mated together. There were 100 flies in the offspring generation.

1 Which of the two alleles for wing length is dominant and which is recessive? Explain your answer.

2 How many of the offspring generation would you expect to have vestigial wings? Use a genetic diagram to explain your answer.

Monohybrid inheritance involving co-dominant alleles

Snapdragons are commonly grown in gardens. Some snapdragon plants produce red flowers and some produce white flowers. Flower colour is controlled by a single gene that has two alleles, one for red flowers and one for white flowers. In this example, the two alleles are co-dominant. Figure 12 shows what happens when a snapdragon with red flowers is crossed with a snapdragon with white flowers.

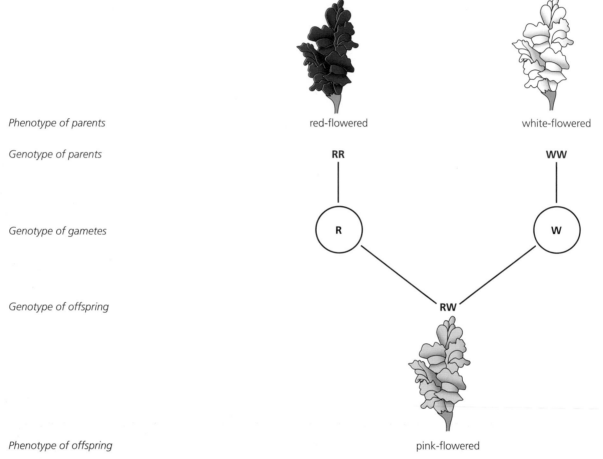

Phenotype of parents	red-flowered	white-flowered
Genotype of parents	**RR**	**WW**
Genotype of gametes	**R**	**W**
Genotype of offspring	**RW**	
Phenotype of offspring	pink-flowered	

▲ **Figure 12** Flower colour in snapdragons is controlled by one gene that has two co-dominant alleles. The offspring of a red-flowered plant and a white-flowered plant all have pink flowers.

When the red-flowered plant is crossed with the white-flowered plant, all the offspring are heterozygous. If the allele for red flowers was dominant, all these heterozygotes would have red flowers; if the allele for white flowers was dominant, all these heterozygotes would have white flowers. As you can see in Figure 12, the heterozygotes produce flowers that are neither red nor white; they are pink. Why are they pink? Because the allele for red flowers and the allele for white flowers are co-dominant – they both show their effect in the phenotype of a heterozygote.

You can see the effect of co-dominant alleles in Figure 13. This diagram shows two different crosses, one between two plants with pink flowers and another between a plant with pink flowers and a plant with white flowers.

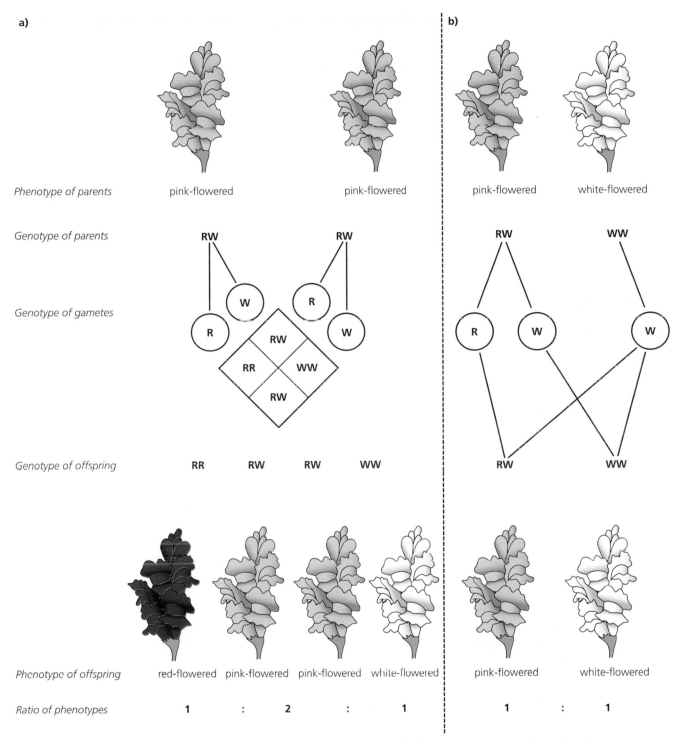

a)

Phenotype of parents	pink-flowered	pink-flowered
Genotype of parents	RW	RW

Genotype of gametes

Genotype of offspring   RR   RW   RW   WW

Phenotype of offspring	red-flowered	pink-flowered	pink-flowered	white-flowered
Ratio of phenotypes	1 :	2 :	1	

b)

pink-flowered	white-flowered
RW	WW

Genotype of offspring   RW   WW

pink-flowered	white-flowered
1 :	1

▲ **Figure 13 a)** Crossing two pink-flowered snapdragon plants, and **b)** crossing a pink-flowered snapdragon with a white-flowered snapdragon.

You can see that, instead of the 3:1 ratio we expected in monohybrid inheritance with a dominant and recessive allele (Figure 11), we get a ratio of 1:2:1 (Figure 13a).

4 Genes, alleles and inheritance

Inheritance of coat colour in horses

Like all mammals, horses have hair. We refer to a horse's hair as its coat. Some horses have a red-coloured coat, called bay, and some have a white-coloured coat, called grey. One gene for coat colour is involved, but it has two alleles, **R** and **r**. If a horse is homozygous **RR**, all the hairs in its coat will be red and the horse will have a red-coloured coat. If a horse is homozygous **rr**, all the hairs in its coat will be white and the horse will have a white-coloured coat. However, if a horse is heterozygous, **Rr**, half the hairs in its coat will be red and half will be white. The result is a horse with a pinkish-colour coat. This coat colour is referred to as red-roan. Coat colour in horses is an example of monohybrid inheritance with co-dominant alleles of a single gene. Figure 14 shows horses with these three coat colours.

1 In the coat of a red-roan horse, half the hairs are red and half are white. What does this suggest about the alleles of the coat-colour gene in the hair-producing cells?

When a stallion with a red-coloured coat was mated with a mare with a white-coloured coat, all the offspring were red-roan.

2 Why, in the above statement about Figure 14, was it unnecessary to explain whether the parent horses were homozygous or heterozygous?

3 What ratio of coat colour would you expect in the offspring of red-roan parents? Use a genetic diagram to explain your answer.

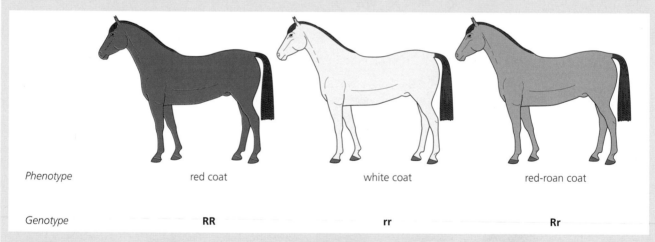

Phenotype	red coat	white coat	red-roan coat
Genotype	**RR**	**rr**	**Rr**

▲ **Figure 14** Horses with a red-coloured coat (bay) are homozygous **RR**; horses with a white-coloured coat (grey) are homozygous **rr**. Half the hairs in the coat of heterozygotes, **Rr**, are red and half are white producing a coat colour called red-roan.

Q6

What ratio of blood groups would you expect among the offspring of a mother with the genotype I^AI^O and a father with the genotype I^BI^O? Explain your answer.

Monohybrid inheritance involving multiple alleles and co-dominance

We have looked at monohybrid inheritance involving genes with only two alleles. Many genes have more than two alleles; we refer to them as **multiple alleles**. We saw in Table 2 (page 106) the genotypes and phenotypes involved in the inheritance of the human ABO blood group. This feature involves multiple alleles (three alleles of one gene) and it involves co-dominance. Figure 15 shows a cross involving the ABO blood group. One parent is blood group AB and the other is blood group O.

Monohybrid inheritance involving multiple alleles and co-dominance

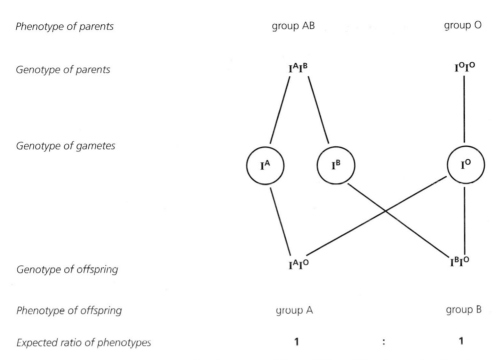

Phenotype of parents	group AB	group O
Genotype of parents	$I^A I^B$	$I^O I^O$
Genotype of gametes	I^A I^B	I^O
Genotype of offspring	$I^A I^O$	$I^B I^O$
Phenotype of offspring	group A	group B
Expected ratio of phenotypes	1 :	1

▲ **Figure 15** A parent of blood group AB and a parent of blood group O will have children of blood group A or blood group B, but none with the same blood group as the parents.

Using the χ^2 test in examining genetic crosses

We can use the χ^2 test when examining data that are categoric. Data from genetic crosses are often categoric, for example data from the inheritance of the human ABO blood group.

Geneticists investigated the ABO blood group phenotype of children born to a large number of parents. For each pair of parents, the blood-group genotype of one was heterozygous $I^A I^O$ and the other was heterozygous $I^B I^O$. The geneticists expected the blood groups of children in the investigation to be A, B, AB and O in a ratio of 1:1:1:1.

1 Explain why the geneticists involved a large number of parents in this investigation.

2 Explain why the geneticists expected a ratio of 1:1:1:1 in the offspring.

Table 3 shows the results the geneticists obtained.

Blood group	Observed (O)	Expected (E)	(O − E)	(O − E)²	$\frac{(O - E)^2}{E}$
A	26				
B	31				
AB	39				
O	24				
				$\chi^2 = \sum \dfrac{(O - E)^2}{E} =$	

Table 3 The results of an investigation into the inheritance of human blood groups.

3 Copy the table and use the prediction of the geneticists to complete the column showing the expected number of children with each blood group (E). Explain how you arrived at your answer.

4 Calculate the values for $(O - E)$, $(O - E)^2$, $\dfrac{(O - E)^2}{E}$ and $\sum\dfrac{(O - E)^2}{E}$. Enter these values in your copy of Table 3.

The geneticists calculated a value of $\chi^2 = 4.46$. They looked up their calculated value of χ^2 in a probability table. Table 4 shows part of this table.

Degrees of freedom	Values of χ^2 at each probability level		
	$p = 0.10$	$p = 0.05$	$p = 0.01$
1	2.71	3.84	6.64
2	4.61	5.99	9.21
3	6.25	7.82	11.34
4	7.78	9.49	13.28

Table 4 Values of χ^2 at three probability levels.

5 How many degrees of freedom were there in the geneticists' data? Explain your answer.

6 What does the calculated value of χ^2 tell us about the results of this investigation? Explain your answer.

7 State a null hypothesis for this analysis.

Inheritance of sex

In humans, the members of a homologous pair of chromosomes are usually the same size and carry the same genes in the same order. This is not, however, true of chromosome pair 23. Figure 16 shows chromosome pair 23 from a human male. One of the chromosomes (the Y chromosome) is very short and carries few genes; the other (the X chromosome) is long and carries many more genes. Although the short Y chromosome carries very few genes, one of them is crucial in determining sex. This gene, called the testis-determining gene (or SRY), turns the developing sex organ of a seven-week-old embryo into a testis, and the embryo develops into a male. Because the X chromosome carries some genes which the Y chromosome does not, a male is effectively haploid for these genes.

A human male has one X chromosome and one Y chromosome; his genotype is **XY** and half his sperm cells will carry an X chromosome and half will carry a Y chromosome. A human female has two X chromosomes; her genotype is **XX** and all the eggs she ever produces will carry an X chromosome.

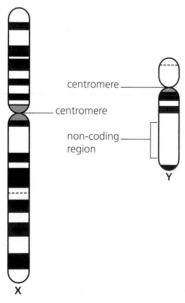

centromere

centromere

non-coding region

Y

X

▲ **Figure 16** Chromosome pair 23 in a human male. The two homologous chromosomes are different sizes and only share a small number of genes. The shorter Y chromosome does not have a copy of many of the genes present in the X chromosome.

Use a genetic diagram to explain why there is a 1 in 2 chance of a child being a girl. Use the symbols X and Y to denote X and Y chromosomes.

Monohybrid inheritance involving a sex-linked character

A **sex-linked character** is one that is controlled by a gene located on one of the sex chromosomes. In most cases, a sex-linked gene is on the X chromosome. Examples of sex-linked characters in humans include haemophilia (a disorder in which the blood of sufferers – called haemophiliacs – clots only slowly), red-green colour blindness and Duchenne muscular dystrophy.

Figure 17 represents what might happen when a woman with blood that clots normally has children with her husband who is a haemophiliac. The symbols we are using in this diagram represent two things: the sex chromosome and the alleles of the gene for blood clotting. The allele for normal blood clotting is dominant, so we represent it as **H**. The allele for slow blood clotting is recessive, so we represent it as **h**. However, this gene is located on the X chromosome. We show this using the symbol X^H for an X chromosome with the **H** allele and X^h for an X chromosome with the **h** allele. Since the Y chromosome has no copy of the gene for blood clotting, it is shown simply as Y.

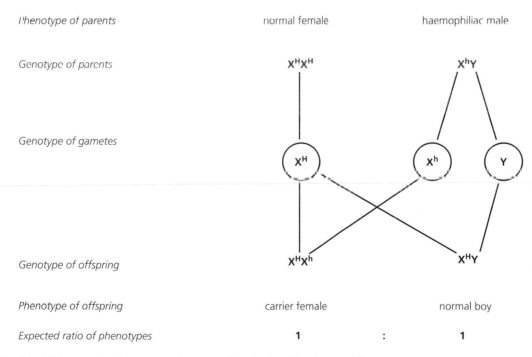

Phenotype of parents	normal female	haemophiliac male
Genotype of parents	$X^H X^H$	$X^h Y$
Genotype of gametes	X^H	X^h Y
Genotype of offspring	$X^H X^h$	$X^H Y$
Phenotype of offspring	carrier female	normal boy
Expected ratio of phenotypes	1 :	1

▲ **Figure 17** Possible children produced by a normal woman whose husband is a haemophiliac.

Notice in Figure 17 that any boy born to this couple will have blood that clots normally; he will not be a haemophiliac like his father. Any girl born to this couple will be heterozygous for the blood-clotting gene. We call her a **carrier** for haemophilia but, since the **H** allele is dominant, her blood clots normally. Figure 18 shows what might happen if this girl who is a carrier grows up, marries a man whose blood clots normally and has children.

4 Genes, alleles and inheritance

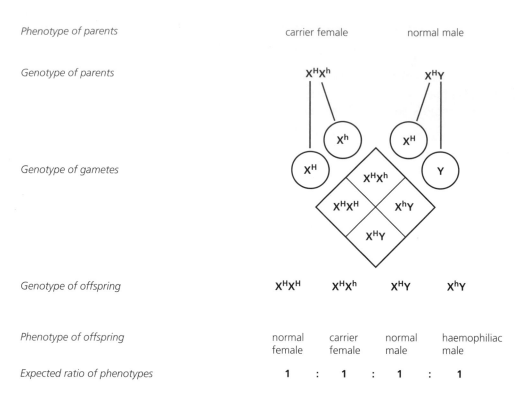

Phenotype of parents		carrier female	normal male	
Genotype of parents		$X^H X^h$	$X^H Y$	
Genotype of gametes				
Genotype of offspring	$X^H X^H$	$X^H X^h$	$X^H Y$	$X^h Y$
Phenotype of offspring	normal female	carrier female	normal male	haemophiliac male
Expected ratio of phenotypes	1 :	1 :	1 :	1

▲ **Figure 18** Possible children produced by a woman who is a carrier for haemophilia and her normal husband.

Q8

Suggest why sex-linked genes are usually on the X chromosome rather than on the Y chromosome.

Q9

Would it be possible for a girl to have haemophilia? Explain your answer.

Notice that the woman in Figure 18 can produce a son whose blood clots normally and a son who has haemophilia. The haemophiliac son inherits the recessive gene for haemophilia on the X chromosome from his mother, whose own father was a haemophiliac (see Figure 17). This is typical of characteristics that are controlled by genes that are located on the X chromosome – the condition appears in males in alternate generations.

Interpreting pedigrees

So far, we have used genetic diagrams to explain a pattern of inheritance. An alternative method for presenting information about inheritance is shown in Figure 19 opposite. This diagram shows a **pedigree**. In pedigrees such as this, females are represented by circles and males by squares. Couples who have children are linked by a horizontal line and the children are represented in a hierarchical fashion below them.

■ Practise using the symbols in Figure 19 by describing the phenotypes of individuals 1 and 2.

Individual 1 is represented by a square, so he is male. The square contains the symbol A, so he has blood group A. Individual 2 is shown by a circle containing the symbol B, so she is a female with blood group B.

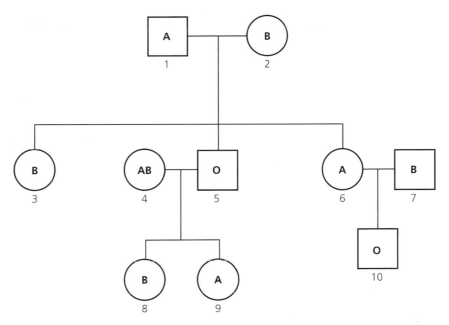

▲ **Figure 19** This pedigree shows the inheritance of ABO blood groups in three generations of a family.

■ **Which of the individuals 3, 4, 5, 6 and 7 are the children of couple 1 and 2?**

The diagram shows a horizontal line between individuals 1 and 2. This indicates that they had children together. The vertical line from individuals 1 and 2 goes directly downwards to connect to each of their children, so individuals 3, 5 and 6 are the children of the parents 1 and 2. Individuals 4 and 7 are not connected by vertical lines to the parents 1 and 2, so are not their children. You can see from the horizontal lines that individual 4 is the partner of 5, and individual 7 is the partner of 6.

A pedigree only tells us phenotypes, it does not tell us genotypes. When we interpret a pedigree, we have to use clues provided by the phenotypes of some individuals to work out the genotypes of as many people in the pedigree as we can. The skill you need to develop is to identify which individuals provide these clues.

■ **Look carefully at Figure 19. Is there any individual whose genotype you can be sure of?**

Individual 5 is blood group O, so we know his genotype is I^OI^O. Having found this, we can work backwards and forwards through the pedigree to work out other genotypes.

■ **Now that we know the genotype of individual 5, we can work out the genotypes of his parents. What are they?**

Since individual 5 has the genotype I^OI^O, he must have inherited one I^O allele from each parent. However, his father (individual 1) has blood group A, so his father's genotype must be I^AI^O. The mother of individual 5 (individual 2) has blood group B, so her genotype must be I^BI^O.

■ **Now that you have worked out the genotypes of individuals 1 and 2, give the genotypes of their other children, individuals 3 and 6.**

Individual 3 is blood group B. She must have inherited the I^B allele from her mother (individual 2). If she had inherited the I^A allele from her father, she would be blood group AB. Since she is blood group B, she must have the genotype I^BI^O. Individual 6 is blood group A. She must have inherited the I^A allele from her father. She cannot have inherited an I^B allele from her mother or she would be blood group AB, so she must have the genotype I^AI^O.

■ **If individuals 6 and 7 were to have another child, what is the probability that it would be another boy with blood group O?**

Since individual 10 is blood group O, we know he must have the genotype I^OI^O. This means that his mother (6) and father (7) must carry the I^O allele, so they are I^AI^O and I^BI^O, respectively. Consequently, they are equally likely to produce offspring with the genotype I^AI^O, I^AI^B, I^BI^O or I^OI^O. If you are not sure why, draw a genetic cross diagram of $I^AI^O \times I^BI^O$ to see what offspring can result. This gives a probability of a child with blood group O of 0.25. The probability of a child being a boy is 0.5. So, the probability of a boy with blood group O is $0.25 \times 0.5 = 0.125$.

■ **Why would '1 in 8' not have been the correct answer to the previous question?**

Notice that the previous question asked for the probability. A probability has a maximum value of 1, a minimum value of 0 and intermediate values that are decimal fractions, for example 0.125. The answer '1 in 8', although it represents the chance of something happening, is not the correct way of expressing the probability.

Chapter 5

Gene pools, selection and speciation

Figure 1 You might have seen our common rabbit species, the European rabbit, feeding in fields or parks early in the morning or during the late evening. It might surprise you to learn that rabbits were introduced by the Normans from mainland Europe about 800 years ago and were originally bred for food and fur. With changes in agricultural practices, rabbits became serious agricultural pests in the 20th century.

In the 20th century, the rabbit population in the UK grew so large that rabbits became a serious agricultural pest. They also became a pest across mainland Europe and in Australia, where they had been introduced to provide a food store for early settlers.

In the late 19th century, laboratories in Uruguay had discovered a virus that caused skin lesions in the local rabbits but which killed European rabbits. This virus was the myxomatosis virus and the disease that killed European rabbits is myxomatosis. During the 1940s, a number of research organisations studied the use of myxomatosis as a potential biological control method to reduce rabbit populations.

Initially, governments were unwilling to sanction this method of biological control. One reason for the UK government's reluctance was public opinion: although farmers wanted to be rid of rabbits, most people in the UK had been brought up on cultural traditions that were sympathetic to rabbits. Perhaps you had a fluffy toy rabbit, or read stories involving rabbit characters, as a child. Another reason was that scientists at the time feared the virus might 'jump the species barrier' and infect animals other than rabbits. If this idea seems familiar, exactly the same concern has been expressed more recently about the avian influenza virus.

Before governments decided to sanction the use of myxomatosis to control rabbits, events took their course. A retired physician, who owned a rabbit-infested farm, released the virus in France in mid-1952. From France, the virus reached England, at Edenbridge in Kent, in 1953 and rapidly spread throughout the UK, killing almost 90% of the rabbit population. In Australia, the virus 'escaped' from a government laboratory in the Murray Valley sometime during the Christmas–New Year holiday in 1950–51. It quickly spread throughout the Murray-Darwin basin, killing millions of rabbits.

Although myxomatosis initially devastated rabbit populations, they have since recovered. Rabbit populations are nothing like as large as they were, but they are now in balance with the myxomatosis virus. How has this balance happened? One reason is that rabbits have become resistant to the myxomatosis virus. This has occurred through a process called natural selection, in which rabbits with genes giving resistance to myxomatosis had greater reproductive success than rabbits that were susceptible to myxomatosis.

It might surprise you to learn that natural selection has also changed the virus population. In England, the myxomatosis virus is transmitted from rabbit to rabbit by fleas, when they bite rabbits to feed on their blood. However, in Australia, the virus is transmitted by mosquitoes. A mosquito will only bite a living rabbit, so any virus which kills its host too quickly is unlikely to be passed on to another host by a mosquito. In Australia, natural selection has favoured myxomatosis viruses that are less virulent, allowing their hosts to live longer and providing greater opportunity for the virus to be passed to another host via a mosquito bite.

Gene pools

Organisms do not normally live alone; they normally live in populations. A **population** is a group of organisms of the same species living in the same habitat. In Chapter 4, we looked at the inheritance of genes by individuals. Now we will look at the genes controlling a single characteristic in a population.

▲ **Figure 2** The banded snail is common in woodlands and grasslands in the UK. The difference in shell colour is controlled by two alleles of a single gene.

Figure 2 shows two adults of a single species of the banded snail. One of the snails has a yellow shell and the other has a pink shell. This shell colour is controlled by two alleles of a single gene. The pink allele (C^P) is dominant over the yellow allele (C^y). Imagine there are 1000 snails in a population living in a beech woodland. Since every snail has two alleles in its genotype for shell colour (C^PC^P, C^PC^y or C^yC^y), we can say that there are 2000 alleles in the shell-colour genotypes of snails in this population. This represents the gene pool for this characteristic in this population. In general, we can define the **gene pool** as the total number of alleles of a particular gene that are present in a population at any given time.

Allele frequencies in the gene pool of a population

We saw above that a population of 1000 individuals of banded snail has a gene pool of 2000 alleles of the gene for shell colour. If all the snails in this population had the genotype C^PC^P, all the alleles in the gene pool would be C^P. In other words, the **frequency** of the C^P allele in this gene pool would be 1. If all the snails had the genotype C^PC^y, the frequency of the C^P allele in the gene pool of this population would be 0.5 and the frequency of the C^y allele would also be 0.5. Note that frequencies are always expressed as decimals, not as vulgar fractions.

The Hardy-Weinberg principle and the Hardy-Weinberg equation

Hardy and Weinberg were two scientists who looked at **allele frequencies** within populations. The principle that carries their names, the **Hardy-Weinberg principle**, predicts that the frequency of the alleles of one gene in a particular population will stay the same from generation to generation. In other words, there will be no genetic change in the population over time. The principle is based on mathematical modelling and makes the following five important assumptions.

- **The population is large.** In small populations, chance events can cause large swings in allele frequencies. You learnt about some of these, such as the founder effect and genetic bottlenecks, in your AS Biology course.
- **There is no movement of organisms into the population (immigration) or out of the population (emigration).** Any such movement of organisms would result in new alleles entering the gene pool or existing alleles leaving the gene pool.
- **There is random mating between individuals in the population.** This ensures that there is an equal probability of any allele of a gene being passed on to the next generation.
- **All genotypes must have the same reproductive success.** This also ensures that there is an equal probability of any allele of a gene being passed on to the next generation.
- **There is no gene mutation.** Any mutation of genes would cause some alleles in the gene pool to change to different alleles of the same gene.

Using the Hardy-Weinberg equation

We cannot always tell the genotype of an individual from its phenotype. For example, you learnt in Chapter 4 that an organism showing the effect of a dominant allele could be homozygous for that allele or it could be heterozygous. This is where the Hardy-Weinberg equation is helpful. It allows us to calculate the frequencies of alleles in the gene pool and of genotypes and phenotypes in the population.

■ For which type of phenotype can you always tell the genotype?

We can always tell which phenotype has the homozygous recessive genotype because the recessive allele only shows its effect in a homozygote. In a heterozygote, the effect of the recessive allele is masked by that of the dominant allele.

The Hardy-Weinberg equation is applied to two alleles of a gene. The symbol **A** represents the dominant allele, which will always show its effect in the phenotype, and the symbol **a** represents the recessive allele. Since we are dealing with allele frequencies, we need symbols to represent the frequencies of the two alleles in the gene pool. We use the following symbols:

- p = frequency of the **A** allele in the gene pool
- q = frequency of the **a** allele in the gene pool.

Q1

Would you expect the Hardy-Weinberg principle to have held true over the past 20 years in the human population of the town in which you live? Explain your answer.

■ Explain why $p + q = 1$.

The gene pool for this gene consists of only two alleles, **A** and **a**. The frequency of the gene itself is 1.0, since all organisms have the gene. Consequently, the frequencies of the two alleles added together must also be 1, i.e. $p + q = 1$.

To summarise so far, we have a population with a gene pool for a particular gene that has two alleles. The frequency of the dominant allele, **A**, is p and the frequency of the recessive allele, **a**, is q.

■ If the frequency of the **A** allele in the population is p, what is the frequency in the population of individuals with the genotype **AA**?

The frequency of the **A** allele is p, so the frequency of the **AA** genotype must be $p \times p$, or p^2. We can work out the frequency in the population of each of the three possible genotypes, **AA**, **Aa** and **aa**. The Hardy-Weinberg equation tells us that, in a single population:

- the frequency of the **AA** genotype is p^2
- the frequency of the **aa** genotype is q^2
- the frequency of the **Aa** genotype is $2pq$.

The equation also tells us that $p^2 + 2pq + q^2 = 1$.

If you do not feel confident about the mathematics, do not worry. So long as you remember the above frequencies, you will find the calculations involving the Hardy-Weinberg equation are very easy sums to do.

Let's go back to our beechwood population of 1000 banded snails. When a group of students investigated this population, they found that 160 of the snails had yellow shells and 840 had pink shells. We always start with the phenotype which gives away its genotype, i.e. the homozygous recessive. In this case, yellow-shelled snails are homozygous recessive (C^yC^y).

■ There were 160 snails with yellow shells in the population of 1000 snails. What was the frequency of yellow-shelled snails in this population?

To work out this frequency, you must remember that frequencies are always given as a decimal value. There were 160 yellow-shelled snails in the population of 1000 snails. The frequency of these yellow-shelled snails is 160/1000 = 0.16.

■ Now we have worked out the frequency of homozygous recessive snails as 0.16, what is the frequency of the recessive allele in this population?

The Hardy-Weinberg equation tells us that the frequency of the homozygous recessive individuals is q^2. So, if we know the frequency, we simply need to use a calculator to find its square root. In this case:

- the frequency of the yellow-shelled snails $q^2 = 0.16$
- therefore, the frequency of the allele for yellow shells $q = \sqrt{0.16} = 0.4$.

■ We can now calculate the frequency of the **A** allele in the gene pool. What is its frequency?

The Hardy-Weinberg equation tells us that $p + q = 1$. We have just calculated the value of q, so we find the value of p as $1 - q = 1 - 0.4 = 0.6$.

If you did these calculations correctly, you can pat yourself on the back. We are now in a position to work out the frequency of the three genotypes controlling snail shell colour in the beechwood population.

- The frequency of the genotype $C^P C^P = p^2 = 0.6 \times 0.6 = 0.36$.
- The frequency of the genotype $C^P C^y = 2pq = 2 \times 0.6 \times 0.4 = 0.48$.
- The frequency of the genotype $C^y C^y = q^2 = 0.4 \times 0.4 = 0.16$.

We know that our calculation must be correct because $p^2 + 2pq + q^2 = 0.36 + 0.48 + 0.16 = 1$.

■ We know from the data above that the students found 160 yellow-shelled snails in the woodland population. How many snails in the population were **a)** homozygous with pink shells; **b)** heterozygous with pink shells?

Hopefully, you will now find the last part of the calculation very easy; all we need to do is to multiply the frequency by the number of snails in the population.

- The frequency of homozygous pink-shelled snails is 0.36. There were 1000 snails in the population, so the number of homozygous pink-shelled snails is $0.36 \times 1000 = 360$.
- The frequency of the heterozygous snails is 0.48, so the number of heterozygous pink-shelled snails is $0.48 \times 1000 = 480$.

We can always check that we are correct, because if we add up the number of different snails we have calculated, we should find they give us the total in the population. In this case, $160 + 360 + 480 = 1000$.

Natural selection resulting from differential reproductive success

The Hardy-Weinberg principle tells us that allele frequencies in a large population remain stable from generation to generation. One of the assumptions on which the Hardy-Weinberg principle is based is that all genotypes in the population have equal reproductive success. This will not be true if organisms with one particular genotype are:

- more likely to die before reproducing
- unable to grow sufficiently well to reproduce successfully
- unable to attract a mate.

In all these cases, organisms with one particular genotype will reproduce less successfully than others in the population with different genotypes and will leave fewer, or no, offspring. We say there is **differential reproductive success** between the genotypes in the population.

Q2

Populations of banded snails also live in grassland, where they are preyed on by song thrushes. Which shell colour would you expect to be more common in grasslands? Explain your answer.

Differential reproductive success is common in populations. In beech woodlands, the yellow-banded snails shown in Figure 2 are very conspicuous against the pink leaf litter lying on the ground. The snails with pink shells are better camouflaged and so are more difficult to find. Song thrushes (Figure 3) eat banded snails. Like us, song thrushes have colour vision. A large number of investigations have shown that song thrushes find more of the conspicuous yellow-shelled snails than they do pink-shelled snails in beech woodlands. As a result, fewer yellow-shelled snails survive to reproduce. This means that fewer of the C^y alleles are passed on to the next generation. The process by which the frequency of alleles in a population changes, because different genotypes have differential reproductive success, is called **natural selection**. Table 1 summarises the process of natural selection.

Figure 3 The song thrush eats banded snails.

Sequence of events leading to natural selection	Application of these events to selection of yellow-banded snails in beech woodlands
Within a population, a characteristic has more than one phenotype. These phenotypes result from genetic variation in the genotypes controlling this characteristic.	The C^P and C^y alleles of the shell colour gene result in snails with pink shells and snails with yellow shells.
There is differential reproductive success between the different phenotypes.	Yellow-shelled snails are more conspicuous than pink-shelled snails among beech litter. Song thrushes find yellow-shelled snails more easily than they find pink-shelled snails. Fewer yellow-shelled snails survive to reproduce.
Organisms with greater reproductive success leave more offspring than those with less reproductive success.	In a beech woodland, pink-shelled snails have more offspring than yellow-shelled snails.
Organisms with greater reproductive success will pass their favourable allele to their offspring. As a result, the frequency of this allele will increase in the population, i.e. natural selection has occurred.	In a beech woodland, the frequency of the pink allele will increase and the frequency of the yellow allele will decrease. Pink-shelled snails are at a selective advantage in beech woodlands.

Table 1 An explanation of natural selection in a beech woodland population of banded snails. The events in the left-hand column can be applied to any example of natural selection.

Directional and stabilising selection

Natural selection operates on most characteristics that organisms possess and results in populations being better adapted to their environment. It can result in a change in a population from one phenotype to another or it can result in a reduction of variation about an optimum modal value.

Directional selection

Directional selection acts against one of the extremes in a range of phenotypes. As a result, one phenotype becomes rare and an alternative phenotype becomes common (Figure 4).

Look first at the upper of the two graphs. It represents a frequency distribution of phenotypes in a population. The graph is a normal distribution with a fairly large standard deviation. The mode (most frequent value) is marked in red. The graph represents the frequency distribution of this population before natural selection has occurred. The lower graph is a frequency distribution of the same population after natural selection has occurred. The standard deviation of this curve is less than the upper curve and its mode has shifted to the right on the x-axis. Natural selection has caused a genetic change in this population, favouring organisms with a characteristic towards the upper range of the frequency distribution.

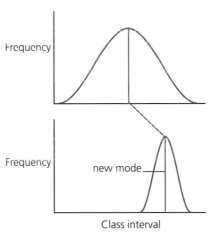

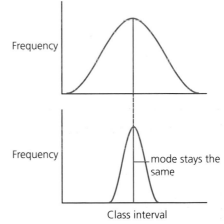

Figure 4 The graphs show variation in one characteristic in a population. The upper graph shows the range of phenotypes before natural selection, the lower graph shows the range of phenotypes after directional selection has occurred.

Figure 5 Stabilising selection reduces the variation of one phenotype in a population. The upper graph shows the range of phenotypes before natural selection, the lower graph shows the range of phenotypes after stabilising selection has occurred.

Q3

How can you tell that the lower curve in Figure 4 has a smaller standard deviation than the upper curve?

Q4

The National Health Service (NHS) was introduced to the UK in July 1948. Suggest why natural selection on human birth mass has had less of an effect since 1948 than before this date.

Stabilising selection

Stabilising selection acts against both extremes in a range of phenotypes. As a result, variation about the mode is reduced (Figure 5). Again, the upper graph shows the frequency distribution of this population before natural selection. The lower graph shows the frequency distribution of the same population after natural selection has occurred. This time, the mode is in the same position: this is the most advantageous phenotype. Stabilising selection has reduced the variation about this modal value.

Stabilising selection occurs on birth mass in humans. Babies with very low or very high birth masses have a higher infant mortality rate than those with a birth mass near the mode of the range.

Natural selection and sickle-cell anaemia

Many large towns and cities in the UK have sickle-cell anaemia clinics. Sickle-cell anaemia is caused by a mutant allele of the gene controlling the production of β-globulin, one of the polypeptides in a haemoglobin molecule (Figure 6). Although the mutant allele changes only one amino acid in the β-globulin chain, its effect is striking.

When the concentration of oxygen is low, haemoglobin molecules with the mutant β-globulin chains have abnormally low solubility and form fibres within red blood cells. This causes the red blood cells to change from disc-shaped cells to sickle-shaped cells. You can see these differences in Figure 7. The sickle shape reduces the surface area of red blood cells.

Sickle cells are also targeted for destruction by the immune system, so have a much shorter life span than normal red cells (about 15 days compared with the normal 120 days). Both these differences result in anaemia. Sickle cells are also liable to get stuck in capillaries, so that nearby tissues become starved of oxygen.

1 The mutant allele changes only one amino acid in the β-globulin chain. It is likely that the mutation involves a base substitution and not a base deletion. Explain why.

2 Sickle cells have a smaller surface area and shorter life span than normal red blood cells. Explain why both these differences result in anaemia.

The production of β-globulin is controlled by a single gene that has two alleles. We will call the β-globulin gene **Hb** and its alleles **Hb^A** and **Hb^S**. Table 2 shows the possible genotypes and phenotypes for this characteristic.

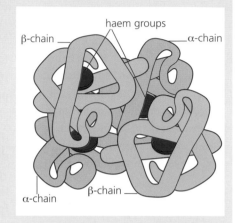

Figure 6 A haemoglobin molecule consists of four polypeptides, two α-globulin chains and two β-globulin chains, and four haem groups that bind to oxygen.

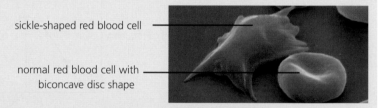

sickle-shaped red blood cell

normal red blood cell with biconcave disc shape

Figure 7 Red blood cells from a person suffering sickle-cell anaemia. Sickle cells have haemoglobin molecules with β-globulin chains that differ by one amino acid from those in normal disc-shaped cells.

Genotype	Phenotype
Hb^{A}Hb^A	All red blood cells have normal haemoglobin and are disc shaped.
Hb^{A}Hb^S	Almost all red blood cells have normal haemoglobin; a few have abnormal haemoglobin and become sickled at low oxygen concentrations. People with this genotype have the sickle-cell trait, but are healthy and usually show no symptoms of sickle-cell anaemia.
Hb^{S}Hb^S	All red blood cells contain abnormal haemoglobin, which causes all red blood cells to become sickled in low oxygen concentrations. People with this genotype have sickle-cell anaemia, the complications of which can shorten their lives.

Table 2 The genotypes and phenotypes associated with sickle-cell anaemia.

3 What does Table 2 tell you about the **Hb^A** and **Hb^S** alleles?

4 Sickle-cell anaemia can shorten people's lives. Explain why you might expect natural selection to reduce the frequency of this allele in human populations.

Despite its apparent disadvantages, sickle-cell anaemia is common among people of African, Middle Eastern or Southern Mediterranean descent. To explain this, we need to look at how malaria is spread. Malaria is common in Africa and in parts of the Middle East and the Southern Mediterranean. Malaria is caused by a single-celled organism, called *Plasmodium falciparum*. This parasite is transmitted when a female *Anopheles* mosquito bites and sucks blood from an infected person and then injects saliva, containing several parasites, as she bites another person. Figure 8 shows a female *Anopheles* mosquito feeding on human blood. Once in the blood system of a human host, the parasites enter the host's red blood cells where they use oxygen within the red blood cells for their own respiration.

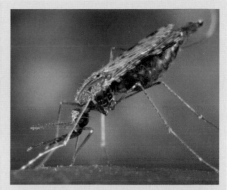

Figure 8 A female *Anopheles* mosquito must have a blood meal before she lays her eggs. She pierces human skin with her sharp mouthparts and sucks blood. If she carries the malarial parasite (*Plasmodium falciparum*) she will infect the human she has bitten.

5 Suggest how the behaviour of a red blood cell that is infected by *P. falciparum* will differ in someone who suffers from sickle-cell anaemia and someone who does not.

The reduction in oxygen concentration caused by respiration of *P. falciparum* causes the red blood cells of a sickle-cell anaemia sufferer to become sickled. This causes them to be destroyed by the body's immune system within about 15 days. As they are destroyed, the *P. falciparum* within them is also destroyed. This makes people with the sickle-cell trait less susceptible to malaria than people with no sickle-cell condition.

6 In an area where malaria is endemic, which genotype, **Hb^AHb^A**, **Hb^AHb^S** or **Hb^SHb^S**, will be at a selective advantage? Explain your answer.

7 What is the probability that two parents, both of whom have the sickle-cell trait, will have a child that suffers from sickle-cell anaemia? Use a genetic diagram to justify your answer.

Q5

Canidae is the name of a taxonomic group, called a family. Name the taxonomic group represented by **a)** *Canis* and **b)** *familiaris*.

Speciation

Figure 9 on the next page shows three closely related species in the taxonomic family *Canidae*. The three dogs belong to different breeds but they are all the same species called *Canis familiaris*. Different breeds of dog originated when humans bred ancestral dogs that had useful characteristics. Jack Russell Terriers (**c**) were originally bred for hunting foxes, Border Collies (**d**) were originally bred for herding sheep and Bullmastiffs (**e**) were originally bred to find and immobilise poachers. When humans breed animals or plants for their useful characteristics, we call this **artificial selection**. The other animals in Figure 9, the wolf (*Canis lupus*) and the jackal (*Canis aureus*), are naturally occurring species. They arose by natural selection. The formation of new species by natural selection is called **speciation**.

You learnt in your AS Biology course that organisms belong to the same **species** if they breed together in their natural habitat and produce fertile young. So how can one species give rise to another during speciation? Some people imagine that a female produces one group of young, some of which belong to a different species from herself. This is possible, but it is thought to occur mainly in plants. For example, a failure of homologous

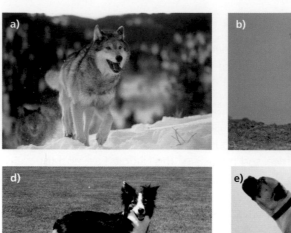

Figure 9 These animals belong to the same taxonomic family, *Canidae*. **a)** is a wolf; **b)** is a jackal; **c)**, **d)** and **e)** are three breeds of dog – **c)** is a Jack Russell Terrier; **d)** is a Border Collie; **e)** is a Bullmastiff. The different species resulted from natural selection and the different breeds of dogs resulted from artificial selection by humans.

Figure 10 This male fruit fly has very short (vestigial) wings, which make it less successful in courtship than males that have long wings.

chromosomes to separate during meiosis in plants can result in offspring that contain multiple copies of the haploid complement of chromosomes. This condition is called **polyploidy**. Polyploid plants are unable to breed successfully with plants that have the normal, diploid complement of chromosomes. In our definition of a species, these polyploid plants belong to a new species that is different from the diploid plants from which they originated. However, this rarely, if ever, happens in animals.

The accumulation of genetic differences is vital for speciation to occur by natural selection. Figure 10 shows a fruit fly, belonging to the genus *Drosophila*. You might have seen fruit flies like this on over-ripe fruit in your home or around your dustbin in summer. There are many different species of fruit fly. The differences between them are so small that you would probably not be able to tell one from another. However, the flies can tell. Before mating, male and female fruit flies undergo a courtship ritual. The male performs a dance, during which he vibrates his wings, changes his body position and licks the female. The whole sequence, including the response of the female, is controlled by many genes and is species-specific. If a male does not have the correct courtship dance, the female does not allow him to mate. The male in Figure 10 is homozygous for a mutant allele of the gene for wing length. As a result, he has very short (vestigial) wings. This will not kill him, but it means that his courtship dance does not attract a female to mate with him. He will not pass on his alleles of the gene for wing length.

As a result of gene mutation, the male fruit fly in Figure 10 will die without leaving any offspring. His mutated genes will not be passed on. However, suppose these genetic changes occurred not in an individual but in an entire population that was isolated from another population of the same species. We now have conditions in which speciation can occur.

Geographic isolation is important in speciation

To understand the importance of geographic isolation in the formation of a new species, let's imagine a population that has become isolated. This might happen, for example, if a pregnant rabbit was washed on a piece of driftwood from the mainland to an offshore island where there were no other rabbits. Here, our female would give birth to a group of offspring. Rabbits do not have the same taboos as humans about mating with their brothers and sisters, so these rabbits would mature and breed with each other. In a short space of time, there would be a population of rabbits on the island. Because rabbits are not strong swimmers, there would be no interbreeding between the rabbits in the island population and those in the mainland population. Not only would the island population be geographically isolated from the mainland population, it would be reproductively isolated as well. We can say that there will be no **gene flow** between the island gene pool and the mainland gene pool.

You learnt in Chapter 4 that gene mutations occur at a constant, low rate in any population. Some gene mutations are beneficial, resulting in organisms with the beneficial alleles of a gene reproducing more successfully than others. As a result, the frequency of these beneficial alleles in the gene pool of this population will increase. In our imaginary example, this process will happen separately in the mainland and island populations of rabbits. Over a period of time, differences in their gene pools will accumulate. Look back to the definition of species that we used earlier. If any of the accumulated differences in the two gene pools prevent rabbits from the mainland and island populations mating then, by our definition, we now have two species of rabbit. If any of the accumulated differences in the two gene pools result in hybrids between the mainland and island populations being infertile then, again according to our definition, we have two species of rabbit.

In summary, geographic isolation can lead to reproductive isolation which, in turn, can lead to the accumulation of differences in the gene pools of two populations of the same species. If these differences prevent mating between individuals from the two populations or result in the infertility of hybrids between them, the two populations have become different species.

Q6

Islands in an archipelago often have species of a single genus that are unique to each island and all different from the species on the mainland. Explain why.

Evolution of copper tolerance

The mine waste which formed the hill shown in Figure 11 contains high concentrations of copper ions. The fields around the copper waste contained populations of bent grass but for many years none grew on the hill.

1 Seeds from bent grass plants are dispersed great distances by the wind. Suggest why no bent grass plants grew on the hill in Figure 11.

After several years, bent grass plants began to grow on the hill. Experiments found that these plants had developed a tolerance to copper in the soil.

2 Suggest how copper tolerance had originated in the copper-tolerant plants.

3 The genes that enabled bent grass to become copper-tolerant spread through the population as a result of natural selection. However, the populations growing in the fields around the hill are not copper-tolerant. Suggest why.

4 Suggest how you could test whether these two populations have become separate species.

Figure 11 This hill was formed when waste was tipped from a nearby copper mine.

Chapter 6
Response

Consider the following range of actions.

- **You eat a meal. Over the next few hours the food moves through your gut, enzymes are produced and digestion occurs.**
- **You become aware of a full bladder. You visit the toilet and at the appropriate moment release the contents of your bladder.**
- **A 'friend' flashes a bright torch beam in your eyes. You blink. Thinking that this is a fun activity, the joker continues to flash the torch. After a few times you manage to stop yourself from blinking.**
- **You start to walk across a field and then notice a large bull watching you. You turn and run, with your heart thumping but at a speed that surprises you.**
- **You are attracted to someone and you decide to ask them on a date. Was this love at first sight? What was it that first attracted you to the person?**
- **You select a particular track on your MP3 player and listen to it. What made you choose this particular track?**

▲ **Figure 1** Is this love at first sight?

Each of the actions above involves one or more responses to one or more stimuli. In some of the actions you have control over your response, or at least awareness and partial control. In the others you have no conscious control and only limited awareness.

After swallowing your food you have no control over the digestive processes. Fortunately, you don't have to remember to release enzymes as soon as the food gets into the stomach and then some more when it moves into the intestines! You don't even realise that this is happening. When your bladder is full you will be able to control when you empty it, unless things get desperate, but you are unlikely to be aware exactly how you exercise this control. Blinking is a reflex action which you are unlikely to prevent when taken by surprise; but you can suppress the action when you are expecting the stimulus.

The other three actions seem to involve your conscious control. The decision to run would be conscious, even though not a particularly good one if the bull could run faster than you. The increase in heart rate and the surge of adrenaline that improves your performance would have occurred automatically. Similarly, the attraction to your intended date may feel as though it is conscious and rational, but there is much evidence that people respond unconsciously to a range of signals from potential partners.

The decision to select a particular piece of music seems to be entirely conscious and deliberate. Imagine that you are sitting lazily listening to your MP3 player. Every so often you tap your finger on the click wheel to move to another song. This is an example of a very simple, basic kind of decision-making: deciding to tap your finger for no particular reason. Nevertheless, it seems that it *is* your conscious decision; you intend to tap your finger, so you do. A sequence of experiments carried out by a Californian neuroscientist called Benjamin Libet seems to suggest otherwise.

What happens in your brain when you decide to tap your finger? It takes time for a brain to do anything, and for a brain to make a body do anything, so what is the sequence of events?

- First, there may be a stimulus, such as an unpleasant sound on the track you are hearing. The signal takes up to 30 ms to reach the relevant part of your brain.

- Next, there is a wave of activity in a part of your brain called the motor cortex. This can be detected experimentally using electrodes applied to the scalp. It is called a readiness potential (RP). It usually lasts for about 500 ms, but in some cases can last up to 800 ms.

- Finally, there is a period of activity in the motor nerves leading from your motor cortex to the muscles that control your finger. These impulses travel very fast, and your finger moves less than 50 ms after the signal has left your brain.

You can see that the finger action can occur approximately 550 ms after the readiness potential first appears in the brain – that is just over half a second later. So when is the moment of conscious decision? Libet's experiments were designed to find out. The surprising result was that we apparently only become aware of our conscious intention to act about 200 ms before the action. Libet interpreted this to mean that the brain sets off the decision-making process unconsciously, about 350 ms before we are consciously aware that we want to act. You can find out more about the evidence on which Libet's conclusions were based in the exercise on the website that supports this book.

Advances in technology have made it possible for neuroscientists to discover much more about, for example, how our brains store memories and enable us to think. Some of this research may help us to understand and treat some of the disorders of mental functioning.

The processes that enable you to see, interpret, understand and hopefully memorise what is written here are very complex. Not surprisingly, we will only be able to introduce some of the basic principles. In this and the next chapter we will study how the nervous system responds to internal and external stimuli and how it coordinates the information it receives. Perhaps some of you may eventually go on to discover more about how our brains operate and what underlies our human experiences.

Survival and response

How does a blackbird find an earthworm in a lawn? How does a cat catch a blackbird? How does a blackbird recognise a cat as a dangerous predator? How does a worm escape from a blackbird? How does a worm find a mate? Getting food, avoiding being eaten, finding favourable conditions to live in and reproducing are all essential requirements for survival. Any species that does not have the ability to respond to these requirements will die out.

▲ **Figure 2** Predator against prey – can the cat beat the birds' survival responses?

Any survival strategy in an animal involves the sequence:

As we shall see, the complexity of behaviour involved varies enormously according to circumstances and from species to species. Detection requires a **stimulus** that can be registered by **receptors**. Receptors are specialised cells that produce electrical activity in nerve cells. For example, a sudden movement by a cat may be the stimulus that is detected by receptors in a bird's eyes. **Processing** involves nerve impulses being transmitted to the brain (or in a worm to a sort of mini-brain) and from there to the parts of the body that will produce the appropriate **response**. The response is carried out by **effectors**. In the bird the effectors will be the muscles that operate the wings. So, the full sequence is:

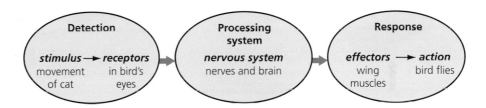

This response obviously has vital survival value to the bird. It is important that the response should be rapid, so a **reflex action** must be involved. At the same time it cannot be an absolutely fixed and automatic action. The escape flight has to be controlled. It must carry the bird away from the cat and travel far enough to reach a safe place. The bird is able to take control and undertake much more complex behaviour than just flapping its wings.

A simple reflex arc

In humans the simplest response to an external stimulus is an **uncontrolled reflex action**. If, for example, you accidentally touch a hotplate on a cooker you will very rapidly pull your hand away. You won't have to think about your action and you won't be able to stop this response if you are not prepared for the heat. This has the obvious advantage that it minimises the damage that might be caused.

Figure 3 on the next page shows the **neurones** (nerve cells) involved in a reflex arc that produces a rapid response. The high temperature of the hotplate stimulates pain receptors close to the surface of the skin. These are actually thin branches at the end of a **sensory neurone** that has a long fibre that extends all the way through the arm to the spinal cord. The receptors produce nerve impulses that are transmitted to the opposite end of the sensory neurone. This end of the fibre of the sensory neurone has tiny branches that almost touch similar branches on a **connector neurone**. A junction between neurones is called a **synapse**. The connector neurone has another synapse linking to a **motor neurone**. The fibre of the motor neurone transmits impulses to a muscle in the arm and stimulates it to contract. This produces the response that causes the arm to be quickly pulled away.

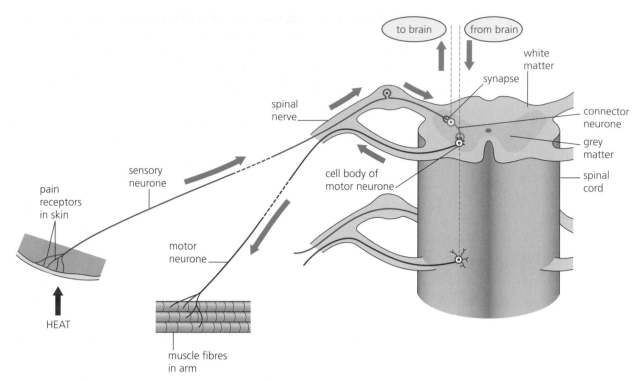

▲ **Figure 3** A reflex arc.

At its simplest the reflex arc involves only three neurones. This is the system that results in the totally automatic reflex response. In practice, however, the response of pulling the arm off a hotplate includes much more than a single reflex arc. Touching the hot surface stimulates not just one but a large number of receptors, so many sensory neurones send impulses to the spinal cord. The response requires the coordinated contraction of several muscles and many muscle fibres in the arm. Therefore the connector neurones have to despatch impulses not just to one muscle but to several, which is why they have synapses connecting with many other neurones. Impulses also pass up to the brain, making us conscious of pain and enabling us to take further action, as well as to shout 'ouch'.

Finding the right environment

The ability to respond to environmental stimuli occurs in almost all living organisms. Even seemingly simple organisms that do not have a complex nervous system can find favourable conditions or get out of trouble. One way is just to keep moving when they experience an undesirable stimulus. This is called **kinesis**, which simply means movement. (As you may recall from the term 'kinetic' energy.) A second, rather more advanced response, is to move directly towards or away from a stimulus. This is called a **taxis**.

Example of kinesis

Planarian worms live in shallow streams and ponds where they are carnivorous scavengers. They belong to one of the most primitive of the invertebrate phyla. They have a network of nerve fibres and simple 'eyes' that have light-sensitive cells but no lens.

Planarian worms are often found clustered on the underside of stones, where they normally remain hidden during daylight. If a stone is turned over, the worms immediately start moving around in random directions. When their movements bring them back into the darkness they stop moving. This behaviour helps to protect them from predators. Laboratory experiments show that the brighter the light the more rapid the worms' movements are. The worms move around randomly until they happen to get to a darker environment, so this is directionless movement, a kinesis.

Example of taxis

Euglena viridis is a single-celled organism that that lives in small ponds. It has chloroplasts (Figures 5 and 6) so it requires light for photosynthesis. It also has a long flagellum that it uses for swimming. There is an area near the base of this flagellum that is sensitive to light. You can also see a red spot close to the light-sensitive area.

Euglena responds to light by swimming towards it. Waves of contraction pass along the flagellum from the base to the tip. These waves pull the *Euglena* forward. Because there is only one flagellum the waves also make the cell rotate as it moves forward. If the light is shining from one side the red spot will shade the light-sensitive area each time the cell rotates. When it is moving straight towards the light the area will stay illuminated all the time. The molecules of the light-sensitive pigments are aligned so that when they are illuminated they stimulate the flagellum to beat. This results in the *Euglena* moving directly towards the light. This response is therefore a taxis and because it is movement towards light it is called a **phototaxis**.

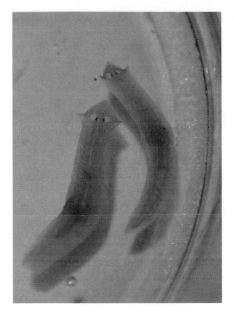

▲ **Figure 4** Planarian worms, which are about 15 mm long.

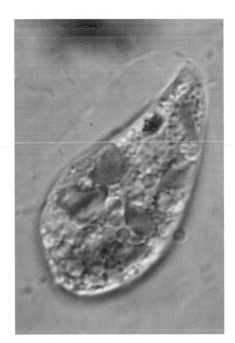

▲ **Figure 5** *Euglena viridis*, magnification × 2500.

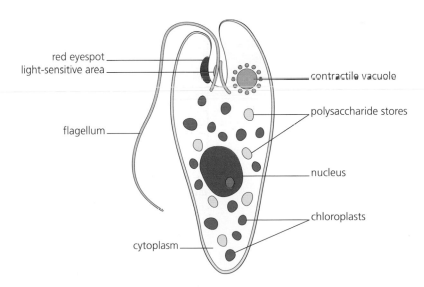

▲ **Figure 6** Structure of *Euglena*.

It is amazing that such seemingly complex behaviour can be coordinated by tiny organelles within a single cell. Nevertheless, the ability to move to and away from light is common in single-celled organisms. In shallow seas, there is a mass migration of vast quantities of plant plankton towards the surface in the daytime and then at night-time back to deeper levels where there are higher concentrations of the mineral nutrients that they require.

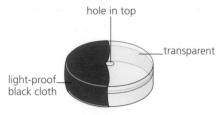

Figure 7 A simple choice chamber can be used to investigate taxes and kineses.

Q1

Identify and explain the features of the student's experimental set-up that helped to control variables other than light.

Q2

Identify and explain weaknesses in the student's investigation and suggest possible improvements.

Q3

Would the student's experiment distinguish whether the response was a kinesis or a taxis? Explain your answer.

Investigating taxes and kineses

It is quite likely that you have already done some behaviour experiments using apparatus similar to the choice chamber shown in Figure 7. These are suitable for studying taxes and kineses in small invertebrates, such as woodlice and maggots.

A student was investigating the hypothesis that woodlice exhibit negative phototaxis – that is they move away from the stimulus of light. The student covered one half of a choice chamber with light-proof black cloth. The other half of the chamber was illuminated by a light bulb fixed about 20 cm above it. Ten woodlice were put into the chamber through the hole in the centre of the top. Five minutes later the student counted how many woodlice could be seen on the illuminated side (and by simple subtraction worked out how many were on the dark side!). This procedure was repeated ten times, using different woodlice each time. Table 1 gives the student's results.

Trial number	Number of woodlice after 5 minutes	
	in light	in dark
1	3	7
2	5	5
3	4	6
4	4	6
5	3	7
6	5	5
7	6	4
8	3	7
9	2	8
10	4	6

Table 1 Results of an experiment to investigate negative phototaxis in woodlice using a choice chamber.

Looking at these results the student concluded that woodlice do move away from light and therefore do show negative phototaxis. The student's conclusion looks reasonable, since there are clearly more woodlice found in the dark than in the light. It also fitted with the student's expectations, since woodlice are normally found in damp, dark places, such as under logs or stones.

However, let's do a statistical test on these results and see if this helps to confirm the conclusion. In the investigation the student was dealing with two discrete categories, light and dark. We can therefore use the chi-squared (χ^2) test to determine whether the result was statistically significant (see page 23). The null hypothesis would be that there is no significant difference between the number of woodlice going to the dark side and the number going to the light side. The equation we need to use is:

$$\chi^2 = \sum \frac{(O - E)^2}{E}$$

Remember that O is the observed results. If we add together the numbers in the light and dark from the ten trials we find that there were 61 woodlice on the dark side and 39 in the light. If the null hypothesis is true we would expect equal numbers on each side, i.e. 50. Table 2 shows the results of our calculations.

	Light side	Dark side
Observed results (O)	39	61
Expected results (E)	50	50
$(O - E)^2$	121	121
$\dfrac{(O - E)^2}{E}$	2.42	2.42

Table 2 Using the chi-squared (χ^2) test to determine whether the results are statistically significant.

Now, taking the sum of the values for the light and dark sides, we find that $\chi^2 = 4.84$.

We now need to look up the values for χ^2 in a table of critical values (see Table 11, page 24). Since there are only two categories there is only one degree of freedom. The critical value for rejecting the null hypothesis is 3.84, so we can have over 95% confidence that there is a significant difference in the number of woodlice moving to the dark. On the other hand, to be over 99% confident the critical value would have to be greater than 6.63.

It is perhaps not surprising that the response of the woodlice is not an absolute movement towards darkness. Woodlice are relatively active organisms that have fairly complex behaviour. There are several species, many of which look much the same to the casual observer, but not all behave in exactly the same way. Most, however, are nocturnal and during the daytime they remain concealed in dark, moist places. At night they may wander considerable distances in search of food sources, and from time to time they do emerge into daylight. Their rate of activity seems to be in part influenced by a 'biological clock', which increases activity during the hours of night-time. A rigid behaviour pattern in which woodlice unfailingly moved away from light would not necessarily always give survival advantage. In practice several factors may interact.

Reactions of woodlice to humidity

Some of the first experiments on woodlouse behaviour were carried out by J. Cloudsley-Thompson. In one series of experiments he studied the response of woodlice to humidity, using large choice chambers similar to the one shown in Figure 8. The humidity on each side was controlled. Under the gauze on one side was a dish of a drying agent that absorbed moisture from the air. On the other side was distilled water that maintained a high level of humidity above it.

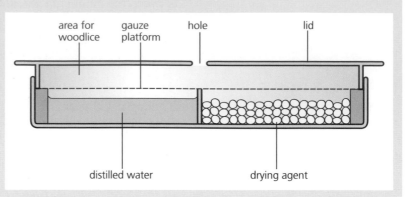

▲ **Figure 8** Apparatus used by Cloudsley-Thompson to investigate woodlouse behaviour.

Cloudsley-Thompson compared three sets of conditions.

A – choice chamber in the light and woodlice that had been in the light for several hours before the experiment.

B – choice chamber in the dark and woodlice that had been in the light for several hours before the experiment.

C – choice chamber in the dark and woodlice that had been in the dark for several days before the experiment.

In each experiment he put five woodlice into the choice chamber and then recorded the positions of the woodlice after 15 minutes (Table 3).

1 What do the results show about the response to humidity?

2 Suggest how this response might be of benefit to the woodlice.

3 Cloudsley-Thompson noticed that after putting the woodlice into the chamber they usually moved around for a short time before becoming stationary. What does this suggest about the type of behaviour shown by the woodlice?

Woodlice position in choice chamber	Number of woodlice		
	A	B	C
Moving around	113	40	17
Stationary on the dry side	21	35	53
Stationary in the central area between the dry and moist sides	49	47	86
Stationary on the moist side	317	378	344

Table 3 The results of Cloudsley-Thompson's three experiments (**A**, **B** and **C**) in which he investigated three different sets of conditions.

4 For each of the experiments **A**, **B** and **C**, calculate the percentage of woodlice that were moving after 15 minutes. Suggest possible explanations for the differences.

5 Research has shown that woodlice respond to light, humidity and touch. They tend to move away from light, towards high humidity and to stop moving when their back is touched. However, these responses are flexible and can be varied according to circumstances. Using the information about the ecology of woodlice given above, explain how a flexible combination of these responses aids their survival.

Do plants respond to stimuli?

Plants may just seem to sit around and do nothing, but in fact their survival is as dependent as that of animals on being able to respond to environmental conditions. Whichever way a seed lands in the soil, the shoot will grow upwards and the roots will grow down. Common observations show examples of responses. A pot plant growing on a window sill will grow towards the light unless it is turned frequently. Many flowers close at night and then open again in the morning. You may have come across so-called sensitive plants that respond to a touch. Lightly brushing against the end of a leaf will stimulate the leaflets to close up like the ripple of a Mexican wave.

▲ **Figure 9** A farmer does not have to make sure his seeds are planted the right way up in the ground.

Some trees can defend themselves when large numbers of caterpillars start munching their leaves by producing nasty-tasting or poisonous substances. This ability can pass from affected to unaffected parts of the tree as soon as one part is attacked. Some trees can pass this message to neighbouring trees, which then produce the same noxious substances before the attackers move in. You might wonder why the trees don't just produce the noxious substances before any caterpillars come on the scene. The answer is probably that the production of the poison uses energy and resources, so it is more economical to wait until the threat of attack is real.

Plants don't have a nervous system, so how are they able to respond to stimuli? Responses such as bending towards or away from light or gravity result from uneven growth. The seedlings in Figure 10 are bending towards the light because they have been stimulated to grow slightly faster on the more shady side of the stem. This sort of growth response to a stimulus from a particular direction is called a **tropism**. Response to light is referred to as **phototropism**, and it may be either positive – towards the light, or negative – away from the light.

A growth response depends on stimulation by chemical substances. Although this is much slower than the electrical activity of nerves it can be surprisingly rapid. For example, the phototropic response of plant shoots can be detected within minutes of exposure to light. We will see how this may work in Chapter 9.

▲ **Figure 10** Increased growth on the shady side of these seedlings results in a positive phototropic response to a light source.

newspaper photograph, which consists of a mass of tiny dots. Where light falls the receptors will be stimulated, and where there is a dark patch there will be no stimulation. Each receptor that is stimulated can pass an impulse to the brain and thus the brain can interpret the pattern. This is more or less how it works at the fovea where all of the receptors are cones. Each cone in the fovea connects through a single bipolar cell and ganglion cell to one fibre in the optic nerve.

The optic nerve has about 1.2 million fibres. Since there are 130 million receptors there obviously cannot be individual connections to the brain for all the receptors. You can see from Figure 19 that each bipolar cell has synapses that connect with several rods. The bipolar cells in turn have more than one synapse with an optic nerve fibre. Although the simplified diagram shows only a few synapses, in practice there must be an average of about 100 receptors with synaptic connections to each optic nerve fibre. The cones outside the fovea may also have several connections. This affects the amount of detail that can be perceived in an image on the retina outside the fovea.

How clearly we can see an image depends on two factors: sensitivity and visual acuity.

- **Sensitivity** is how much light is needed to stimulate the receptors. Rods are stimulated in much dimmer light than cones, so rods are more sensitive.
- **Visual acuity** is how far apart two spots of light can be seen as being separate. This is affected by the way in which the receptors are connected to the optic nerve fibres.

Q7

How many spots of light will be seen in Figure 21c? Explain your answer.

A ray of light that falls on just one cone in the fovea will show up as a spot of light, as long as it is bright enough to stimulate the cone. This is because the cone is connected to a single fibre in the optic nerve via one bipolar cell (Figure 21a). If another ray of light falls on another cone, as in Figure 21b, the brain will interpret this as two separate spots.

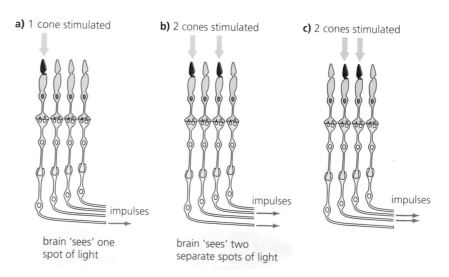

a) 1 cone stimulated   b) 2 cones stimulated   c) 2 cones stimulated

impulses   impulses   impulses

brain 'sees' one spot of light   brain 'sees' two separate spots of light

▲ **Figure 21** How the cones function to allow the brain to distinguish separate spots of light.

Q8

How would you expect the visual acuity of the rods to compare with that of the cones? Explain your answer.

Now consider what will be seen when a ray of light falls on a single rod. Figure 19 on page 147 shows three rods with synapses to each bipolar cell and six rods that connect via the bipolar cells to a single ganglion cell. Assuming that the stimulation of one rod cell is sufficient to send an impulse to the brain, there is no way in which the brain can interpret which of the six rods had been stimulated. In reality there may be many more than six rods with synapses to a single ganglion cell.

So, what is the advantage of having many rods connecting to a single fibre in the optic nerve? As we shall see in Chapter 7, synapses can act as barriers to the transmission of impulses. Although the rod is very sensitive to light, a single rod is unlikely to produce a sufficiently large generator potential to be able to stimulate the bipolar cell to transmit a nerve impulse. However, if a group of rods is stimulated by dim light at the same time, the combined generator potentials will reach the threshold required for transmission across the synapse to the bipolar cell and the fibre in the optic nerve. This means that an impulse can be generated in the bipolar cell and then transmitted to the optic nerve fibre. This process is called **summation**, because the effect on several cells is added together. Although the image is less sharp, we are able to see in much dimmer light than would be possible with cones alone. The eyes of most nocturnal mammals have only, or mostly, rods. For an animal that is hunting or hunted at night it is more useful to see a slightly fuzzy image than to be unable to see at all.

Chapter 7
Coordination

The human brain is estimated to have about 100 billion (10^{11}) neurones. If we gave equal portions of a single brain to every person in England, he or she would get about 2000 cells. Moreover, each neurone has synapses connecting it to as many as 10 000 other neurones. This creates a neural network that makes the average computer seem like a child's toy.

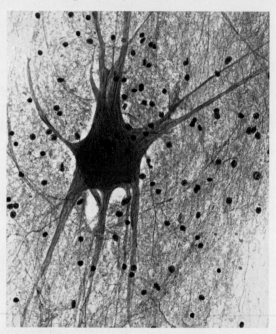

▲ **Figure 1** This photograph was taken by a scanning electron microscope and shows a single neurone and some of its branches that connect it to other neurones in the brain.

The brain is the organ responsible for coordinating the great majority of activities in the body. It synchronises most of the automatic activities such as heartbeat and breathing, as well as such complex movements as walking or playing the piano. The brain receives and processes a constant flow of information from sensory receptors. This information may stimulate appropriate responses, or be stored or ignored. The brain is also responsible for what we consider to be the higher human activities, such as thinking, emotions, memory and consciousness.

Different areas of the brain are responsible for different functions, as you can see from Figure 2, but neuroscientists have discovered that these subdivisions are not sharply defined and that the key to the

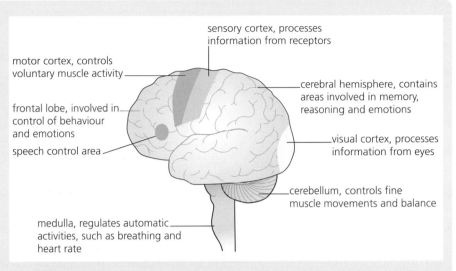

motor cortex, controls
voluntary muscle activity

sensory cortex, processes
information from receptors

frontal lobe, involved in
control of behaviour
and emotions

cerebral hemisphere, contains
areas involved in memory,
reasoning and emotions

speech control area

visual cortex, processes
information from eyes

cerebellum, controls fine
muscle movements and balance

medulla, regulates automatic
activities, such as breathing and
heart rate

▲ **Figure 2** Side view of a human brain. The approximate positions of some parts and their functions are shown. Many other functions take place in central areas that cannot be seen from the outside, or involve coordination between different areas.

brain's functioning lies in its network of connections. It may seem impossible to understand exactly how such a complex organ works, and our progress has been slow. Much of the understanding has come from evidence based on what happens when things go wrong, for example as a result of damage caused by accidents, blood clots and the effects of drugs.

One of the first and most famous examples was an accident that happened to Phineas Gage in 1848. 26-year-old Phineas was in charge of a gang of men building a railway in the north of the USA. The men would drill a deep hole into a rock face, half fill it with explosive powder and then add a fuse and sand. Phineas had to compress the explosive mixture with a long iron bar before lighting the fuse. One day, when not paying full attention, he started to ram down the mixture with his iron bar before sand had been added. A spark caused the powder to explode with such force that the metre long iron bar shot through his left cheek and the front part of his brain before passing out of the top of his skull and landing some 25 metres away. Amazingly, after a few minutes he was able to get up and talk intelligibly. Somehow the doctor who looked after him managed to avoid the wound becoming seriously infected, and after a couple of months Phineas recovered almost fully. He had lost none of his powers of speech or mobility. However, his personality changed significantly. He was no longer the careful, responsible foreman. He became impulsive, unreliable, selfish, dishonest and prone to swearing. As a result he was unable to go back to his old job. He lived for another 12 years but never recovered his previous personality.

This case demonstrated that the frontal lobes of the brain are not essential for basic functions, such as muscle activity, vision or speech, but that they are involved in determining a person's behaviour. It is now known that the frontal lobes are important in organising social behaviour and in planning complex operations.

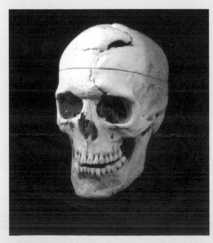

▲ **Figure 3** Phineas Gage's skull clearly showing his injury.

Further observations have helped to clarify what other parts of the brain do. One common observation is that someone who is knocked unconscious, for example in a road accident, has no memory of the events leading up to the accident. This suggests that there is a delay before memory traces are fixed in the brain. It has now been discovered that damage to the hippocampus, near to the base of the brain, prevents the formation of new short-term memories. However, memories of past events, people and places are not affected. It seems that memories are normally first formed in the hippocampus and then later transferred to the cortex of the brain for permanent storage. This may be a way in which the brain 'tidies up' its filing system and 'weeds out' items that do not need to be retained for a long time. The hippocampus is often found to have degenerated in elderly people who are suffering from Alzheimer's disease, which may explain why they have difficulty in remembering day-to-day events but often still retain memories from the distant past.

Other evidence from individuals with localised brain damage suggests that memory and other brain functions depend on complex processes of communication within the brain. In one example a brain-damaged man shown a picture of a rhinoceros was able to give details such as its size and that it lives in Africa, but when asked to describe a rhinoceros he could only say that it is an animal. In another brain-damaged man the opposite was found. He was able to give details when asked to describe a rhinoceros but when shown a picture he was unable to recognise or name the animal or give any information about it. Many similar examples indicate that the brain processes information obtained by sight and by hearing in separate parts. These separate stores then have to be coordinated. Damage to one or other of the stores, or to the system that connects the stores, can destroy the ability to link the two sources of information.

Chemical substances also can interfere with the working of the brain. Alcohol is well known for its mind-altering capacity, as are many other drugs. Alcohol, or ethanol to be more chemically precise, interferes with the synapses. The effect is to inhibit activity in the brain by preventing neurones communicating chemically with each other. For example it inhibits activity in the cerebellum, which normally coordinates fine motor skills and balance. Hence quite small amounts of alcohol make you less steady on your feet and less quick to react. Oddly, alcohol is noted for reducing inhibitions and making people more sociable. This is because it also reduces the neuronal activity that normally inhibits social interactions, making it more likely that someone will reveal personal secrets and do embarrassing things.

So, how is it possible that tiny pulses of electrical activity and the transfer of chemicals across synapses and membranes can result in the complex mass of thoughts, feelings, actions, memories and emotions of which humans are capable? Neuroscientists are still a long way from full understanding, but slowly they are discovering the functions of different regions of the brain and how these interact. In this chapter we shall study the basic processes necessary to understand how the brain works and how the nervous system coordinates our activities. This will enable us to explain how chemicals such as alcohol and other drugs can disrupt the system.

Neurones

First we need to look at the structure of the nerve cells, called **neurones**, which form the conducting tissue in the nervous system (Figures 4 and 5).

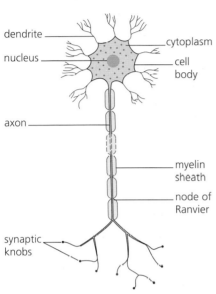

▲ **Figure 4** The structure of a motor neurone.

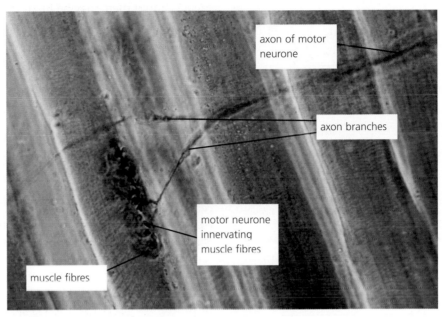

▲ **Figure 5** Light micrograph of a motor neurone and its effector, the muscle fibres.

In Chapter 6 (page 134) you learnt that a simple reflex arc has three different types of neurone.

- A **sensory neurone** transmits impulses to the spinal cord.
- A **motor neurone** conveys impulses from the spinal cord to a muscle.
- One or more **connector neurones** act as links between the sensory and motor neurones.

Figure 4 shows the structure of a motor neurone. You can see that the cell body contains cytoplasm and a nucleus and looks similar in structure to other animal cells. The cytoplasm also contains mitochondria and ribosomes.

In other respects a motor neurone is highly **specialised**. Its cell body is situated inside the spinal cord. It has large numbers of branching fibres, called **dendrites**. A motor neurone may have over a hundred of these thin branches, each with one or more synapses from a connector neurone. There is also one very long fibre that extends from the spinal cord to a muscle. This **axon** can be as much as a metre in length and less than a micrometre in diameter. At the far end of the axon are short branches that terminate in tiny **synaptic knobs**. These knobs fit into little pockets on the surface of muscle fibres. Large numbers of axons are bundled together into **nerves**. The sciatic nerve, for example, originates from the spinal cord in the lower back. Branches from it pass all the way down the leg to muscles in the foot. The sciatic nerve also contains the fibres of sensory neurones conducting impulses in the opposite direction. Damage to the lower back can compress this nerve where it passes between the vertebrae, causing a painful condition called sciatica.

Q_1

In which direction do impulses pass through a motor neurone? Explain your answer.

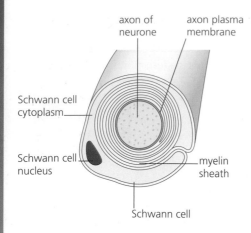

axon of neurone

axon plasma membrane

Schwann cell cytoplasm

Schwann cell nucleus

myelin sheath

Schwann cell

▲ **Figure 6** A section across an axon and Schwann cell.

Q2

Explain why polypeptides and proteins can only be synthesised in the cell body.

Q3

Suggest why it is unlikely that the polypeptides and proteins reach the synaptic knobs by a process of simple diffusion.

You can see from Figure 4 on the previous page that the axon is surrounded by the **myelin sheath**. This sheath is not strictly part of the neurone. It is made from highly specialised cells, called **Schwann cells**, that lie alongside the axon (Figure 6). As the axon grows these cells wrap round and round the axon until there may be up to a hundred layers of lipid and protein membranes. This makes a fatty 'bandage' around the axon shielding it from surrounding tissue fluid. Between each Schwann cell is a tiny gap where the axon is exposed, called a **node of Ranvier**. These nodes are the only places that ions can pass between the tissue fluid and the axon through the plasma membrane. The nodes play an important part in speeding up the passage of impulses along the axon, as we shall see later.

Axons contain no organelles other than a few mitochondria. To stimulate the muscles to contract the synaptic knobs secrete substances into the synapses. To make these substances they need proteins and polypeptides as well as other compounds. These have to be transported all the way from the cell body to the end of the axon. If you imagine the cell body of a neurone being magnified to the size of a golf ball, this is like sending supplies for about a mile down a tube roughly the diameter of a ball-point pen.

Impulses

Impulses are waves of electrical activity passing through a neurone. The process is not the same as the conduction of electricity through a wire. It is much slower, although it is still pretty quick. When a neurone is stimulated at a synapse, there is a brief change in the plasma membrane of the fibre. This allows ions to pass through rapidly. As soon as this happens it causes the next small section of the membrane to change, so ions can pass through here. Meanwhile the first section of fibre changes back as the distribution of ions is restored. This process carries on down the fibre. It is rather like a long line of standing dominoes that are knocked down one after another once the first one is pushed over, except that in this case the domino is immediately set upright again as soon as the next one falls over.

Note that we have referred to a *fibre* and not an axon in this section. This is because in a neurone the term 'axon' should only be used for the long fibre that leads away from the cell body towards a muscle, or other effector. The synapses where a motor neurone is stimulated would be on the dendrites, which are also fibres, but not axons. In fact, impulses pass along all parts of the plasma membrane of a neurone in the same way. This includes the surface of the cell body as well as the dendrites and axon.

So, what causes the movement of ions? Before we explain, it might be a good idea to remind yourself of the ways in which ions can pass through a plasma membrane. In the AS course you learnt about the ways that ions can pass through channels in the membrane. It may be helpful to refer to pages 48 and 49 in the AS book.

The next section guides you through the processes involved in the movement of nerve impulses. Follow each stage carefully and think about the questions in blue before you look at the answers.

■ **Describe the difference between diffusion and active transport of ions.**

In diffusion and active transport protein carrier molecules in the membrane are used. In diffusion the ions pass through protein channels from a higher to a lower concentration by random movement. No energy from respiration is required. In active transport the protein carriers change shape to transport the ions. This requires energy from respiration and it makes it possible to move ions against a concentration gradient.

The resting potential

Look at Table 1 which shows the concentration of sodium, potassium and chloride ions inside and outside the axon of a motor neurone.

Ion	Concentration/mmol dm^{-3}	
	inside axon	outside axon
Sodium (Na$^+$)	15.0	150.0
Potassium (K$^+$)	150.0	5.5
Chloride (Cl$^-$)	9.0	125.0

Table 1 The concentration of sodium, potassium and chloride ions inside and outside the axon of a motor neurone.

■ **Describe the differences in concentration inside and outside the axon for each of the ions in Table 1.**

The concentration of sodium ions outside the axon is much greater than their concentration inside, whereas the opposite is true for potassium ions. The concentration of the negatively charged chloride ions is much greater outside.

■ **There are protein channels in the membrane that allow ions to diffuse through. These channels are always open. From Table 1, what would you expect to happen to the concentrations of the sodium and potassium ions inside and outside the axon?**

You would expect the concentrations to be equalised by sodium ions moving in and potassium ions moving out.

In practice the concentrations are not equalised. Sodium and potassium ions do continuously diffuse through the open protein channels in the axon membrane. However, the potassium ions diffuse through more rapidly than the sodium ions.

■ **This gives the outside of the membrane a more positive charge than the inside. Explain why.**

Sodium ions diffuse in because the concentration inside the axon is low. Potassium ions diffuse out. Because the potassium ions diffuse out faster than the sodium ions diffuse in, more positive ions accumulate on the outside.

If this process kept going the concentrations of both sodium and potassium ions would eventually balance out. However, the difference is maintained by a **sodium-potassium** pump. This pump uses special **carrier proteins** in the membrane to move sodium ions out of the axon and potassium ions in. Three sodium ions are pumped out for every two potassium ions pumped in. The pump uses ATP as a source of energy.

When no impulse is being transmitted, the pump keeps the balance between sodium and potassium ions at the concentrations shown in Table 1. The rate at which the pump works is regulated by the concentration of sodium ions outside the membrane. The result is that the outside of the axon membrane always has a slight excess of positive ions and the inside a slight excess of negative ions. This gives a **potential difference** of about 70 mV (millivolts). Because the inside of the axon is more negative and an electric current is a flow of negative electrons from a more negative to a more positive potential this is written as −70 mV. This is called the **resting potential**.

■ **Which method of ion movement transfers the sodium and potassium ions through the membrane? Explain your answer.**

The sodium-potassium pump uses active transport. It uses energy from respiration to transport the ions and it moves them against their concentration gradients.

Look again at Table 1 on page 155. You will see that the concentration of negative chloride ions is also unbalanced. These also could be expected to diffuse into the axon. However, the movement of ions does not only depend on the concentrations on each side of the membrane. Because the ions are charged, it also depends on the balance of charge on either side of the membrane. What the table does not show is that the cytoplasm inside the axon contains many negatively charged protein molecules, whereas the tissue fluid outside does not.

■ **How do you expect the presence of the protein molecules to affect the diffusion of chloride ions?**

The protein molecules are too large to pass out through the membrane. They give the inside of the axon a high negative charge. This inhibits the diffusion of negative chloride ions into the axon because the negative charges cannot be equalised by the protein molecules moving out.

■ **The cytoplasm of an axon contains mitochondria. Suggest why these mitochondria are important.**

The sodium-potassium pump needs a constant supply of energy in the form of ATP from respiration.

The action potential

So far we have only considered an axon that is not transmitting an impulse. What happens when an impulse passes along an axon?

You may recall from the AS course that some channel proteins in membranes have **gates** which prevent diffusion when they are closed. The diagram from the AS book is repeated in Figure 7 here. Gated **ion channels** play an important role in the neurones. Gates can be opened either by a small change in voltage (**voltage-gated** channels) or by a chemical substance becoming attached to the protein (**chemically-gated** channels). Don't imagine that these channels have a little door that swings open like a garden gate! The channel proteins in fact change shape in such a way that the ions are able to diffuse through.

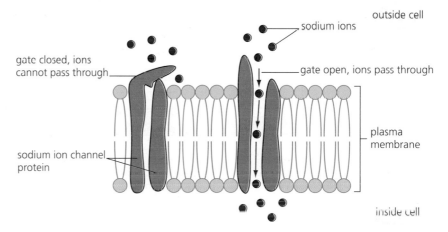

▲ **Figure 7** Gated sodium ion channels in the plasma membrane.

When an axon is inactive the charges are in an unbalanced state all the time. There is an excess of negative charges inside the axon. As we have explained previously, this is the resting potential. Energy is required to maintain this uneven balance. However, this resting potential means that the axon is poised to 'fire' at any time, because of the steep concentration gradient of ions across the membrane.

Stimulation of a neurone at a synapse causes a small change in the potential difference in the membrane close to the synapse. When the change in the potential difference reaches a **threshold value**, the gates in the voltage-gated sodium ion channels change their shape to 'open' for a brief period. This allows sodium ions to rush in. It is estimated that each open channel allows about 20 000 Na$^+$ ions to diffuse through. This causes a sudden increase in positively charged ions inside the neurone. Instead of having a slight excess of negative ions, the inner surface of the membrane now acquires a positive charge of about 30 mV compared to the outside. The change in the potential difference across the membrane is called **depolarisation**. Depolarisation of a small section of a nerve fibre creates an **action potential**.

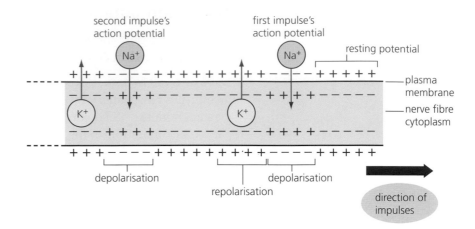

▲ **Figure 8** A change in charge at a nerve fibre membrane during the passage of impulses.

The action potential in one small section of the membrane sets off depolarisation of the next section of the nerve fibre because the change in voltage in the membrane stimulates sodium ion channels to open. Therefore the action potential spreads like a wave along the nerve fibre. This wave is what we call a **nerve impulse**. Once started the impulse travels all the way to the next synapse.

■ **You will see from Figure 8 that the section of the axon behind the first impulse has returned to the resting potential, with the outside being positively charged again. Suggest why it is important that this happens quickly.**

So that another impulse can follow as soon as possible.

The start of depolarisation causes voltage-gated potassium ion channels in the membrane to open, allowing potassium ions to flow out. They, however, flow out more slowly than the sodium ions flow in. The sodium ion channels close as soon as the membrane potential reaches about $+30\,mV$. At this point depolarisation is complete and the inflow of sodium ions stops. The resting potential is restored because potassium ions continue to flow out until the potential difference returns to $-70\,mV$. The nerve fibre is then **repolarised**. In practice, rather more potassium ions flow out so the drop in potential difference overshoots a bit, as you can see from Figure 9. This is called **hyperpolarisation**.

■ **Repolarisation does not immediately restore the concentration of ions inside the axon to their original state. Explain why not.**

The potential difference across the membrane is back to the same value as the resting potential because the balance of positive and negative ions has been restored. However, the balance of potassium and sodium ions has not been restored. There are now more sodium ions and fewer potassium ions inside the axon than before.

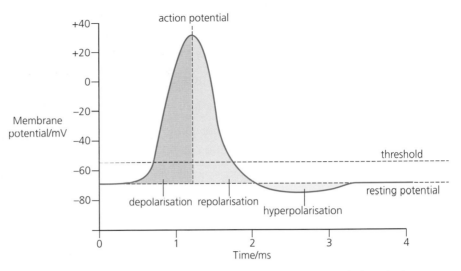

▲ **Figure 9** Changes in membrane potential as an impulse passes a particular point in a nerve fibre.

■ Suggest how the balance of sodium and potassium ions is restored.

The sodium potassium pump restores the balance by active transport. As explained on page 156, this pumps out sodium ions at a faster rate than potassium ions are pumped in. The pump works continuously, so the concentration of positive ions outside the membrane is always rather greater than inside. This maintains the resting potential and keeps the axon ready for another impulse. It is worth noting that the actual proportion of ions that move in and out during the passage of an impulse is tiny. Don't get the impression that all the sodium ions zoom into the axon, or all the potassium ions flow out.

Analysing an action potential

Figure 10 shows the changes in the permeability of the membrane of an axon to sodium and potassium ions as an impulse passes a particular point. The graph also shows the changes in potential difference compared with the resting potential. These results were obtained from the axon of a squid, which has large nerve fibres where it is easier to take measurements than in a mammalian nerve fibre.

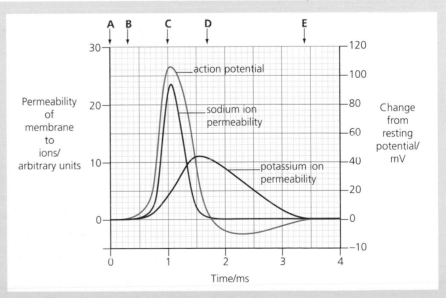

▲ **Figure 10** Graph of changes in permeability to ions of a nerve fibre membrane during an action potential.

1 Use the letters shown in Figure 10 to identify each of the following stages: resting potential, depolarisation, repolarisation, hyperpolarisation.

2 Explain how the change in permeability to sodium ions results in an action potential.

3 Why does the increase in sodium ion permeability start later than the change from the resting potential?

4 Explain the effect of the change in permeability to potassium ions.

5 Look at the curves on the graph showing the changes in the permeability of the membrane to sodium and potassium ions. How do these changes compare?

6 Suggest the advantage of differences in the permeability to sodium and potassium ions.

7 The resting potential in this axon was −70 mV. What was the maximum value of the potential inside the axon?

8 For how long was the potential above −70 mV?

9 For how long did hyperpolarisation last?

How do impulses move along a nerve fibre?

Most neurones are stimulated to produce an impulse at a synapse. For example in a simple reflex arc (Chapter 6, page 134) there is a synapse between the sensory neurone and the connector neurone and another between the connector neurone and the motor neurone. In motor neurones synapses may either be on the branching fibres (dendrites) (Figure 4, page 153) or directly on to the cell body. In sensory neurones there may either be a synapse that links with a receptor cell, as in the eye, or the end of a sensory neurone fibre may be stimulated directly, as in a Pacinian corpuscle (Chapter 6, page 144).

When a nerve fibre is stimulated at a synapse some sodium ion channels open and sodium ions leak in. At first this may be fairly slow depending on the nature of the stimulation. It may be too slow to set off an impulse because the change in membrane potential does not reach the threshold value. But if the increase in sodium ions inside the neurone causes a rise in the membrane potential of 15 mV, the very rapid increase in the opening of the gated sodium ion channels described earlier occurs.

It is this rapid inflow of sodium ions that produces an action potential. Once the rise in potential difference reaches the threshold value an action potential is produced. This always sets off an action potential in the next section of the nerve fibre, so the impulse passes all the way through the neurone. The action potentials have the same potential difference for the full distance, however long the fibre is. This is known as the **all-or-nothing principle**. This means that there are no strong or weak impulses. Like the falling dominoes in a chain, the wave will pass at the same rate right to the end of the line. It also means that the strength of the stimulus does not affect the impulse, as long as the stimulus is strong enough to get above the threshold value. This is rather like squeezing the trigger on a rifle. As long as the trigger is pulled back far enough the bullet is fired. Pulling harder will not make the bullet go any further or faster.

Q4

For a neurone that can transmit 300 impulses per second, how long is the refractory period?

Q5

Look at the graph in Figure 10. At what stage does this suggest that the refractory period ends?

Q6

Suggest why a refractory period is important for nerve function.

There is always a gap between impulses. They never merge together. This is because after the action potential has reached its peak the ionic balance has to be restored by repolarisation. Only after repolarisation can another action potential be generated. The minimum interval between action potentials, and therefore between impulses, is the **refractory period**.

When the gated sodium ion channels close at the peak of the action potential there is a short period of about 0.5 ms when it is impossible for the channels to re-open. Therefore no stimulus can generate an impulse during this period. The full refractory period lasts until repolarisation is complete. In practice a strong stimulus may be sufficient to produce another action potential before repolarisation is complete. The action potential produced is smaller than normal but will produce an impulse that continues along the fibre. Action potentials cannot be generated at such short intervals for long because more and more sodium and potassium ions are on the wrong side of the membrane and the resting potential is not restored. In practice most neurones can only transmit about 300 impulses per second.

The myelin sheath

You may have noticed that so far we have talked about a nerve fibre or axon as though the plasma membrane has a continuous and unobstructed outer boundary. But if you look back to Figure 4 you will recall that in a motor neurone the axon is myelinated. It is surrounded by a fatty myelin sheath that only has gaps at the nodes of Ranvier. Not all neurones in humans have myelin sheaths. For example, many of the neurones in the autonomic system that controls unconscious activities are non-myelinated. **Non-myelinated neurones** conduct impulses in the way described above.

In **myelinated neurones**, ions can only pass through the plasma membrane at the nodes. Therefore action potentials can only occur at the nodes, so impulses in effect jump from node to node, as shown in Figure 11. This is called **saltatory conduction**. (Saltatory is derived from the Latin word for 'dancing' – it has nothing to do with salt.) At each node there is a concentration of sodium and potassium ion gated channels. Since the nodes are about 1 mm apart, saltatory conduction has the great advantage of being much faster. Conduction of impulses in myelinated neurones is approximately 50 times faster than in non-myelinated neurones.

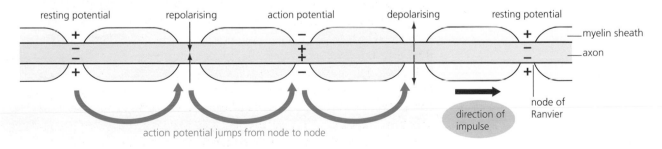

▲ **Figure 11** Saltatory conduction in myelinated neurones. Conduction of impulses in these neurones is approximately 50 times faster than in non-myelinated neurones.

Q7

Suggest why a low temperature reduces the rate of conduction of impulses.

Two other factors affect the rate at which nerve fibres conduct impulses – temperature and the diameter of the fibre. The rate of conduction is slowest at low temperatures and in thin fibres.

In mammals temperature rarely makes a significant difference since they normally maintain a fairly constant temperature. In animals whose temperature varies with the environmental temperature cold conditions can considerably slow reaction times.

The diameter of a fibre makes a difference because thicker fibres can maintain the resting potential more efficiently; a lower proportion of ions leak in or out by diffusion. The myelin sheath in mammals reduces leakage significantly. The fibres can be quite thin and still have a high rate of conduction. This gives mammals an advantage since impulses can be rapidly transmitted long distances, so large animals can respond quickly.

Invertebrates possess only non-myelinated fibres. These are usually quite thin and relatively slow conductors. However, in some circumstances where speed of conduction is particularly vital, some invertebrates have evolved exceptionally large diameter fibres. Much of the early research on nerve conduction was done using giant axons from squids. We saw some of the results in Figure 10. These giant axons have a diameter of nearly 1 mm, over a hundred times greater than other axons in the squid (Figure 12). They can conduct impulses about ten times as fast as normal motor neurones.

The giant axons extend the full length of the main part of the body, with branches to the muscles along the way. When the squid is startled, impulses are conducted at about 35 metres per second along the giant axons. The muscles contract almost instantly, providing the sudden force that squirts water out backwards and jet-propels the squid away from danger. Such violent contraction would not be possible if the impulses travelled slowly along the nerves with the muscles contracting in sequence as the impulses pass. Squids are not the only invertebrates to have a similar adaptation for an escape mechanism. Earthworms for example have large diameter axons running the length of their body. These allow them to contract muscles rapidly and withdraw quickly into a burrow when threatened by a bird.

▲ **Figure 12** Common squid. The main part of the body (on the left) has large muscles surrounding a water-filled cavity. When the squid needs to escape from a predator these muscles contract very rapidly and the squid shoots off using a form of jet propulsion.

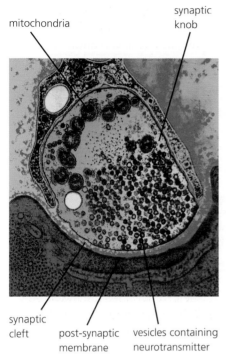

mitochondria

synaptic knob

synaptic cleft

post-synaptic membrane

vesicles containing neurotransmitter

▲ **Figure 13** Coloured scanning electron micrograph of a synaptic cleft. The synaptic knob containing vesicles of neurotransmitter is clearly visible.

Synapses

Synapses are the junctions between neurones. Neurones do not actually touch each other. There is always a small gap called the **synaptic cleft**, which is usually about 20 nm wide. (A nanometre (nm) is a thousandth of a micrometre (μm) and a millionth of a millimetre.) This gap prevents electrical impulses passing directly from one neurone to another. (Note that you should be careful not to refer to impulses crossing synapses.) Instead, communication between neurones is by chemical **neurotransmitters** that cross the cleft from one neurone to another. This is a slower process than the passage of an impulse along a nerve fibre. Figure 13 shows a synaptic cleft.

A synapse lies between the tiny branching end of a fibre and either a dendrite or the cell body of the next neurone. The branching end is always slightly swollen into a **synaptic knob** (Figure 14). A motor neurone may have approximately 8000 synapses on its dendrites and another 2000 directly on the cell body. It has been estimated that neurones in the brain have an average of approximately 40 000 synapses and that some have as many as 200 000. The number of possible pathways between neurones is phenomenal, which helps to explain the amazing complexity of the brain (see the chapter introduction, page 150).

So, what happens when an impulse reaches a synapse at the end of an axon? We will describe here what happens in a **cholinergic** synapse, which uses **acetylcholine** as a neurotransmitter. Acetylcholine (ACh) is the main neurotransmitter in synapses in the nerves outside the central nervous system, as well as in some parts of the brain. It is also used at the junctions between motor neurones and muscles. These **neuromuscular junctions** work in much the same way as synapses, as we shall see in Chapter 8. Many other neurotransmitters are used in the nervous system, but the principles of synaptic transmission are similar in all cases.

When an action potential arrives at the **pre-synaptic membrane** at the end of a synaptic knob, it stimulates gated calcium ion channels to open. The concentration of calcium ions (Ca^{2+}) in the fluid of the synaptic cleft is

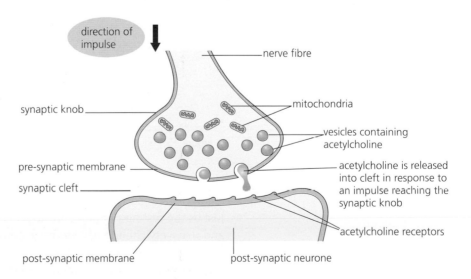

▲ **Figure 14** Structure of a synapse.

higher than the concentration inside the synaptic knob. Therefore calcium ions rapidly diffuse through the open channels into the knob. As you can see from Figure 14, inside the knob there are many vesicles containing acetylcholine. These vesicles are tiny droplets of acetylcholine surrounded by a membrane. When calcium ions enter, some of these vesicles move to the membrane of the cleft and fuse with it. The acetylcholine spills out into the cleft and rapidly diffuses across to the membrane on the other side. This **post-synaptic membrane** has receptor molecules on its surface to which the acetylcholine molecules attach. These receptors are protein molecules. (Don't confuse them with receptor cells, such as the rods and cones that we studied in Chapter 6.)

The receptors are gated channels that are opened chemically rather than by a change in voltage. When acetylcholine molecules attach to the receptor molecules the receptors change shape and allow sodium and potassium ions through. As in other parts of the neurone membrane the sodium ions rush in more rapidly than the potassium ions diffuse out. You will recall that the excess of sodium ions entering the neurone depolarises the membrane. If enough sodium ions enter and the threshold is exceeded, an action potential is set up (page 157). Once an action potential does start it travels all the way along the neurone as an impulse until it reaches the synapses at the other end. Figure 15 illustrates the sequence of the events that occur at a synapse.

1 Action potential opens gated channels. Ca²⁺ ions flow in.

2 ACh vesicles move to membrane and fuse with it.

3 ACh released into synaptic cleft and diffuses across it.

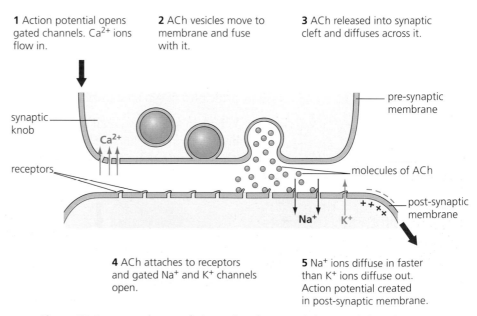

4 ACh attaches to receptors and gated Na⁺ and K⁺ channels open.

5 Na⁺ ions diffuse in faster than K⁺ ions diffuse out. Action potential created in post-synaptic membrane.

▲ **Figure 15** Sequence of events during an impulse transmission at a cholinergic synapse.

You will no doubt have realised that during transmission across a synapse changes take place that must be reversed if it is to keep on transmitting. Just as the resting potential has to be restored in an axon, the ionic balance and the acetylcholine have to be restored in the synapse. Calcium ions are removed from the synaptic knob by active transport, ensuring that the concentration outside the membrane is always higher. The acetylcholine is removed from the receptors by an enzyme called **acetylcholinesterase** that

Q8

You will notice that the synaptic knob contains several mitochondria. Suggest the functions of these energy-producing mitochondria.

breaks it down to acetate and choline. These diffuse back into the synaptic knob where they are synthesised into acetylcholine once again. Vesicles are refilled, so the acetylcholine is continuously recycled.

The role of synapses

It takes much longer to describe what happens at a synapse than for the transmission to occur. But synapses do slow down the rate at which information passes through the nervous system. There is a delay of about 0.5 ms at each synapse. This is because the rate of diffusion is much slower than the rate at which a nerve impulse is propagated along a nerve fibre, especially by saltatory conduction.

This does not mean that synapses are a bad thing. Synapses are vitally important as the points in the nervous system where the passage of impulses is controlled. Remember the all-or-nothing principle. If impulses could cross straight from one neurone to another without any control, stimulation at any point in the system would spread to all neurones. Specialisation of functions in different parts would be impossible. Some of the control mechanisms involved are complex and there is still much to learn about them, especially about the processes involved in the brain.

One simple control method is that transmission across a synapse can only be in one direction. Only the synaptic knob contains the vesicles of neurotransmitter. Therefore neurotransmitter can only be released into the cleft from the pre-synaptic membrane. Once an action potential is established in the post-synaptic neurone it can only travel as an impulse in one direction because depolarisation always occurs in front of the action potential.

Summation

Impulses arriving at a synapse do not always result in impulses being generated in the next neurone. The electron micrograph in Figure 16 shows the synapses on the cell body of a motor neurone. As you can see there are many synaptic knobs all linking to this motor neurone.

A single impulse arriving at one synaptic knob is not likely to generate an action potential. It may release only one or a small number of vesicles of acetylcholine. Only a few of the gated ion channels in the post-synaptic membrane are opened, so not enough sodium ions pass through the membrane to reach the threshold value. The acetylcholine only remains attached to the receptors for a very brief time and the acetylcholinesterase breaks it down within a couple of milliseconds. If several impulses reach the synaptic knob in quick succession, enough acetylcholine is released for the membrane potential to reach the threshold value. This causes the voltage-gated sodium ion channels to open and produces an action potential. Because the effects of several impulses are added together in a short time this process is called **temporal summation** (Figure 17).

A second way in which an action potential can be generated is when several impulses arrive together at different synaptic knobs on the same cell body. The cell body may then be depolarised enough for an action potential to be generated in the axon. This is called **spatial summation**.

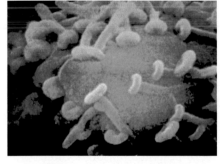

▲ **Figure 16** An electron micrograph showing a mass of synaptic knobs spread over the surface of the cell body of a neurone.

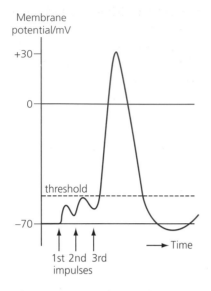

▲ **Figure 17** Temporal summation. You can see from the graph that the first impulse reaching the synapse fails to raise the membrane potential above the threshold for an action potential. The three impulses arriving in quick succession do produce an action potential.

Q9

Which type of summation occurs in the retina? Explain your answer.

Q10

What is the advantage of summation in the retina?

One advantage of summation is that the effect of a stimulus can be magnified. We have, for example, already come across summation in the retina (page 149). Another effect of summation is that it avoids the system being swamped by impulses. A synapse acts as a barrier that only allows impulses to pass on if there is a significant input from receptors or other neurones.

Inhibition

So far we have only described situations in which transmission across a synapse results in an action potential being generated in the post-synaptic neurone. Just as important are synapses where impulses are stopped altogether. Some neurotransmitters prevent action potentials being generated in the post-synaptic neurone. This is called **inhibition**. One way in which inhibition occurs is that the neurotransmitter stimulates gated potassium ion channels to open. Potassium ions therefore flow out of the cell body. If a cell body has both excitatory and inhibitory synaptic knobs attached to it, the effect of sodium ions entering at an excitatory synapse may be cancelled out by an inhibitory synapse. Therefore the membrane potential does not reach the threshold for an action potential to be produced.

Inhibition is very important in **neural circuits**. It enables specific pathways to be stimulated, while preventing random stimulation all over the body. For example in reflex actions it is necessary for neural pathways to produce a response only where it is needed. If your hand is on a hotplate you need your arm muscles to pull it away, but not your leg muscles to contract. The development of inhibitory pathways can be very important in learning specific skills, such as drawing and writing. If you watch a young child learning to draw you will see that at first the movements are all over the place. As the skill develops the movements become much more coordinated as the appropriate pathways are refined by inhibitory circuits.

The effects of drugs

We mentioned in the introduction to this chapter that chemicals may have major effects on the working of the brain and other parts of the nervous system. Many of these substances are chemically quite simple. We often refer to these substances as **drugs**. The term drugs, however, is quite vague and is applied both to substances that are used as cures and to mind-altering substances that may be damaging or potential poisons. Studies of drugs and poisons have revealed that many have quite specific effects on transmission at synapses or on neural circuits within the brain. In some cases research has helped scientists to understand more about how the brain works and to offer ways of using certain drugs to treat mental disorders.

An example of a poison that affects synaptic transmission is **curare**. Curare is a naturally occurring poison that is found in the bark of trees in the Amazon forest. It has been used for centuries by South American Indians to coat the tips of arrows used for hunting. Curare molecules attach to the acetylcholine receptors on post-synaptic membranes, especially the membranes of neuromuscular junctions at the end of motor neurones. The

Q11

Suggest what feature of curare molecules makes them attach to the acetylcholine receptors.

Q12

Suggest how inhibiting acetylcholinesterase affects synaptic transmission.

Q13

One cause of rapid death from organophosphate poisoning is the effect on the intercostal muscles. Explain why.

Q14

Although the heart is muscular it is less affected by organophosphates than the intercostal muscles. Suggest an explanation for this.

curare molecules block the receptors and prevent acetylcholine attaching to the receptors. Therefore impulses that pass down the motor neurones fail to stimulate the muscle contraction, so prey animals are effectively paralysed. Similar substances have been synthesised that can be used to keep muscles relaxed during surgery. This avoids the possibility of sudden muscle convulsions if, for example, a nerve is accidentally stimulated.

Organophosphate insecticides

Other substances that interfere with synapses include **organophosphate insecticides**. These inhibit the action of acetylcholinesterase.

It is estimated that world-wide as many as 300 000 people die each year from organophosphate insecticide poisoning and many more suffer long-term damage to the nervous system. The use of these insecticides is strictly regulated in the UK, but in many developing countries legal requirements are much more lax and often the potential dangers are not appreciated.

Compounds similar to the organophosphate insecticides have been developed into much more potent forms for military use as nerve gases. Because they are readily inhaled these can incapacitate large numbers within minutes. Fortunately most countries have restrained from using them. (You might like to consider the ethical arguments for and against the research efforts that have gone into the development of nerve agents and their antidotes.)

Drugs have a wide variety of effects on the nervous system. We have already seen that at synapses, they may block receptors or inhibit enzymes and prevent re-uptake. Some stimulate the release of a neurotransmitter or have a similar structure that mimics a neurotransmitter. Other possibilities are that a drug blocks movement of ions through gated channels or that it inhibits the sodium-potassium pump. In each case you should be able to work out the effect of the drug on the nervous system.

Electrical synapses

We have so far only described synapses in which transmission is by chemicals crossing a narrow gap between neurones. There are actually some places in the human nervous system where impulses can cross directly from one neurone to another, although these are very much the exception. One example is in the retina of the eye where electrical synapses occur between the bipolar cells and the ganglion cells. If you look back to the diagram of the retina in the previous chapter (Figure 19, page 147), you will see that the bipolar cells lie between the receptors and the fibres of the optic nerve. Electrical synapses have a very narrow cleft. Action potentials arriving at the cleft can directly stimulate an action potential in the next neurone.

■ What difference would you expect in the rate of transmission across an electrical synapse compared with one that used a chemical neurotransmitter? Suggest the advantage of having electrical rather than chemical synapses in the retina.

> The rate of transmission is much faster. There is no delay due to the slow rate of diffusion of the neurotransmitter. This has the advantage of speeding up the transmission of impulses from the receptors in the eye to the brain, which may be important in responding to danger. Since all impulses are passing directly to the brain there are no alternative routes to be controlled.

Other neurotransmitters

Acetylcholine is the neurotransmitter at all neuromuscular junctions in the skeletal muscles, so it is important for all voluntary movement. It is also used by some of the neurones in the autonomic nervous system that controls unconscious activities, such as the heartbeat. In Chapter 6 (page 140) you learnt that the rate of heartbeat is regulated by the sinoatrial node (SAN), which has two sets of nerves connected to it. These sets are called the sympathetic and parasympathetic nerves. The parasympathetic neurones produce acetylcholine at the terminals in the SAN. These neurones slow down the activity of the SAN. The sympathetic nerves increase its activity. Their neurones produce a different neurotransmitter called noradrenaline. This is chemically very similar to the hormone adrenaline, which is involved in the body's response to danger – the so-called 'fight or flight' hormone.

■ Use this information and your knowledge of the action of the SAN to explain how the rate of heartbeat responds to differing demands.

> When the heart needs to increase output, for example during exercise, the sympathetic neurones produce noradrenaline in the SAN. This stimulates the SAN to increase the rate at which the SAN emits the waves of electrical activity that cause first the atria to contract and then the ventricles to pump blood round the body. The cardiac output is increased. When exercise ceases the parasympathetic neurones are stimulated. These produce acetylcholine, which lowers the rate at which the SAN emits the waves of electrical activity, thus restoring the rate of heartbeat to its normal level.

The brain has many different neurotransmitters. Over 50 have been identified. All work in a similar way to acetylcholine. Many occur only in certain parts of the brain and are therefore restricted to particular functions. Some are inhibitory and some excitatory. Each uses receptor molecules that are specific to the neurotransmitter. For example dopamine is a neurotransmitter that is important in parts of the brain concerned with control of muscle action. Parkinson's disease is a fairly common condition in elderly people in whom the neurones that produce the dopamine degenerate. The muscles become too tense, making delicate muscular movements difficult so that, for example, it is hard to pour a cup of tea without trembling. As the disease develops, walking and other everyday activities become harder. Two types of drug can be used in treatment. One is a substance from which dopamine is synthesised in the neurones. Another is an agonist that mimics the effect of dopamine.

■ **Suggest how each of these drugs helps to treat Parkinson's disease.**

Providing the brain with more of the precursor for dopamine helps the neurones that have not degenerated to manufacture more dopamine. The agonist attaches to the same receptors and has the same effect as dopamine.

Another condition in which research suggests that dopamine plays a part is schizophrenia. This is a condition that affects about 1% of people at some time in their lives but most frequently as young adults. Schizophrenia causes people to behave untypically and often to suffer from delusions and hallucinations. It appears to have a variety of causes but one feature that has been discovered is an excess of dopamine receptors in the brain.

■ **Suggest what type of drugs might help to reduce the symptoms of schizophrenia.**

Drugs that block the dopamine receptors should help.

It is not easy to develop drugs that have specific effects in the brain because many substances are blocked from entering the brain by the blood-brain barrier. However, a number of drugs have been produced that help in schizophrenia by blocking the dopamine receptors. Dosage is very important and there is, as with many drugs, a danger of side-effects.

■ **Suggest a side-effect that could result from blocking dopamine receptors.**

A possible side-effect is loss of muscle control, as found in Parkinson's disease.

Neural circuits

We explained on page 166 that synapses can be either excitatory or inhibitory. Combinations of these can produce many different connections between neurones producing complex neural circuits. Large numbers of neurones can synapse with the cell body of a neurone. Whether an action potential is generated in the axon depends on the combination of stimuli received by the cell body.

Look at Figure 18 overleaf. This shows in simplified form the cell body of a neurone and an axon with two synapses on it. Both are excitatory synapses.

If an action potential arrives at synapse **A** the membrane potential becomes more positive but not enough to reach the threshold value for an action potential. This increase in the membrane potential of the cell body is called an **excitatory post-synaptic potential** (**EPSP**). An EPSP on the cell body membrane remains for longer than a change in potential on an axon and it is able to spread along the membrane.

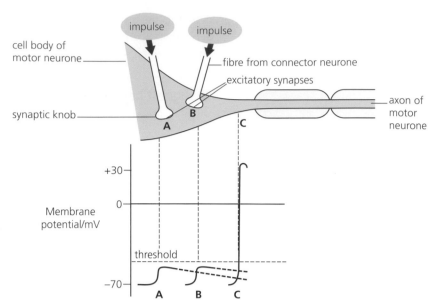

▲ **Figure 18** The cell body and an axon of a neurone with two excitatory synapses on it and a graph of the membrane potential that results.

$Q15$

In the example in Figure 18, impulses from two neurones are combined. What is this effect called?

$Q16$

Suggest what happens at the post-synaptic membrane of the inhibitory synapse to cause the decrease in potential.

If an action potential arrives at synapse **B** at more or less the same time, the two effects are added together and the combined EPSP is now above the threshold value for an action potential to be generated in the axon. As long as the EPSP that spreads to the start of the axon at **C** is above the threshold value an action potential will be set up in the axon.

Figure 19 shows a similar set up, but in this case synapse **B** is an inhibitory synapse. At an inhibitory synapse the effect is that the membrane potential actually decreases. This is not surprisingly referred to as an **IPSP**. In this example the IPSP balances out the EPSP so the threshold value for an action potential is not reached.

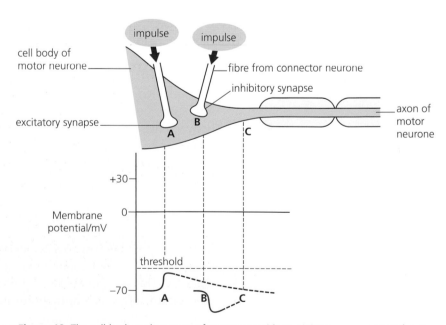

▲ **Figure 19** The cell body and an axon of a neurone with one inhibitory synapse and one excitatory synapse on it. The threshold value for an action potential is not reached.

Chapter 8

Muscles and movement

Have you ever wondered why the meat of chicken breast is almost white and seemingly more or less bloodless, whereas the meat on a chicken leg is much darker and looks more like proper meat? After all, both are muscle. The obvious suggestion might be that farmed chickens can't fly, so they don't need to use their breast muscles for flying as a wild bird does. This, however, is not the complete answer. Chickens have been bred from wild junglefowl that live in the forests of South-east Asia. These birds spend most of their time scratching around for food on the ground, but when danger threatens they quickly fly to safety in a tree. Unlike most wild birds they never fly long distances. The breast muscles that flap their wings are only used in short bursts but have to produce enough force to lift quite a heavy bird off the ground at speed. On the other hand their leg muscles are used more or less all the time for wandering about the forest floor, scratching away at the leaf litter to find food and stopping them falling off their perch.

The junglefowl is adapted to its way of life by having two different types of muscle fibres. The wing muscles mainly consist of **fast-twitch fibres**. These fibres contract rapidly but quickly become tired. They are well suited to a fast and short escape flight into a tree, lasting just a few seconds, but would not be suitable for sustained flight over a long distance. The muscle looks pale for two reasons. It has fewer capillaries than red muscle and therefore obtains its supplies of oxygen and glucose

▲ **Figure 1** Junglefowl need to be able to make short flights to the safety of a tree branch when danger threatens.

▲ **Figure 3** Inheritance of desired muscle fibre types and training methods can make an athlete more successful at their chosen discipline. **a)** For a short race at maximum speed the sprinter benefits from a high proportion of fast-twitch fibres in his leg muscles. **b)** It is an advantage for a long-distance runner to have a high proportion of slow-twitch muscles that fatigue less easily.

Understanding how muscles work is of great importance, not just for sport training but also for other activities, such as the efficient operation of machinery. In this chapter we will study how muscles contract and how they obtain energy needed for contraction. We will also see how their activity is coordinated by the nervous system.

Skeletal muscles

The human body has over 600 skeletal muscles. About 40% of the mass of an average man is muscle. Figure 4 shows some of the muscles on the front (anterior) of the body. You can see how complex the muscular system is – different muscles cross each other in many directions.

You will probably be familiar with the biceps and the triceps muscles in the arm from your GCSE course. These are used to bend the arm at the elbow and lift the lower arm. But as you can see there are several other muscles that enable us to raise and twist our arms as well as making delicate movements with our hands. The biceps and triceps are attached to bones by thick **tendons** made of strong fibrous connective tissue. A tendon has to be flexible, but not elastic. It must not stretch when a muscle is contracting and pulling on a bone. You can see in Figure 4 the tendon that attaches the lower end of the biceps to the radius bone in the lower arm. The upper end of the biceps has two tendons attaching it to the shoulder blade. These cannot be seen in the diagram as the deltoid muscle hides them.

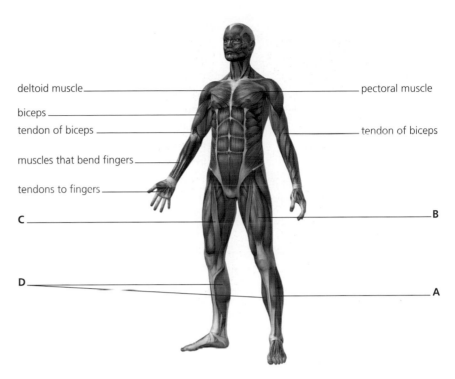

deltoid muscle ———————— pectoral muscle

biceps ————————

tendon of biceps ———————— tendon of biceps

muscles that bend fingers ————————

tendons to fingers ————————

C ———————— B

D ————————

———————— A

▲ **Figure 4** Some of the anterior muscles in the human body.

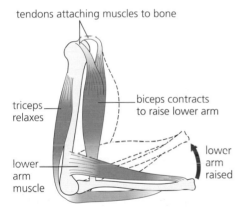

tendons attaching muscles to bone

triceps relaxes

biceps contracts to raise lower arm

lower arm muscle

lower arm raised

▲ **Figure 5** The triceps and biceps in the arm work antagonistically; here the lower arm is being raised.

Q1

Look at muscle **A** in Figure 4. What movement occurs when this muscle contracts?

Q2

Which of the other muscles in the leg labelled **B** to **D** is antagonistic to muscle **A** in Figure 4? Explain your answer.

Some muscles have very long tendons. The muscles that bend the fingers are situated in the arm just below the elbow. Their tendons stretch across the wrist and the palm of the hand. There are also muscles that attach directly to a bone instead of having tendons. For example, if you look at the pectoral muscle in Figure 4 you will see that it is attached all the way down the breastbone.

Muscles only move parts of the body when they **contract**. Muscles can only *pull*; they cannot push. For this reason muscles in general operate as pairs, with one pulling in one direction at a joint and a partner that pulls the opposite way. This is called **antagonistic** muscle action. The biceps and triceps act as an antagonistic pair of muscles (Figure 5). When the biceps contracts the lower arm is pulled upwards. The triceps is relaxed at this time. To straighten the arm again the triceps contracts and the biceps relaxes. Since we usually need more strength to bend the arm, such as when lifting things, the biceps is the larger and more developed muscle.

In practice, when you lift a heavy object in your hand, many muscles are involved. For example, another muscle connects from the lower end of the bone in the upper arm to the lower arm near the wrist. Several muscles are used to enable the hand to grip and the wrist to rotate. Others help to keep the shoulder steady. As the biceps contracts the triceps slowly relaxes, so the movement takes place smoothly. This involves the brain, and in particular the cerebellum, in a highly complex process of coordination. Receptors in the muscles, tendons and joints constantly transmit impulses to the brain indicating the degree of stretching of muscles and the positions of tendons and joints. Reflex adjustments ensure that the operation is a continuously smooth process. Damage to the cerebellum, for example a stroke or tumour, can result in clumsy and jerky movements.

Q3

The intercostal muscles have no tendons. Which bones do they connect and what is their function?

Q4

Describe the ways in which a muscle fibre differs from a typical cell.

The structure of a muscle

The 'cells' that make up skeletal muscles are so different from most cells that they are more sensibly referred to as **muscle fibres**. Not surprisingly, the fibres are long, on average about 30 mm, but in some cases they can be as much as 300 mm. Each fibre is surrounded by a membrane, called the **sarcolemma**. As well as a plasma membrane the sarcolemma has an outer protein layer that separates it from neighbouring fibres and allows each fibre to contract independently. Beneath the sarcolemma is cytoplasm containing many mitochondria. There is also a specialised network of tubules that store calcium ions. The full name for this network is the sarcoplasmic reticulum; for simplicity we shall refer to it as the **reticulum** in this chapter. We shall see later how the reticulum is important in muscle contraction. Dotted along the length of the fibre are many nuclei. Most of the fibre is filled with much thinner fibres, called **myofibrils** (Figure 6).

The muscle fibres are packed together in bundles surrounded by a sheath of connective tissue. It is these bundles that can get stuck in your teeth when eating tough meat. Each bundle is well supplied with blood capillaries and branches of motor neurone fibres. In larger muscles, such as the biceps, many bundles are wrapped together in a thick and tough connective tissue layer. This connective tissue is continuous with the tendon that attaches it to the bone (Figure 7).

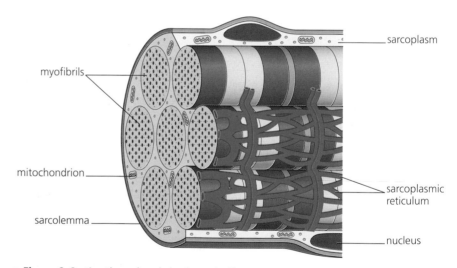

▲ **Figure 6** Section through a skeletal muscle fibre.

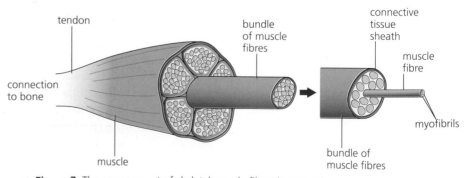

▲ **Figure 7** The arrangement of skeletal muscle fibres in a muscle.

Unit 5 Control in cells and in organisms

Q5

Use Figure 8 b) to calculate the approximate diameter of a myofibril.

Myofibrils

With a light microscope very little detail of muscle fibres can be seen. Dark bands are just visible, as you can see in Figure 8 a), and for this reason skeletal muscle is often called **striated muscle**. An electron microscope enables the structure of the myofibrils inside a muscle fibre to be seen. Figure 8 b) shows an electron micrograph of part of a muscle fibre showing the short sections of five myofibrils.

a) b)

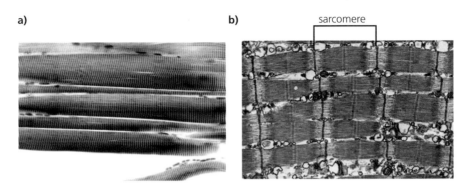

▲ **Figure 8 a)** Skeletal muscle fibres viewed through a light microscope, magnification approximately ×600; **b)** myofibrils in a skeletal muscle fibre viewed with an electron microscope (false colour), magnification approximately ×10 000.

Look carefully at Figure 8 b). You will see that each myofibril has dark (blue) bands across it. The section between a pair of dark bands is a **sarcomere**. In the centre of each sarcomere is a paler (blue) band, and to either side of this are broad darkish (green) areas. You will notice that there appear to be many thin lines running along the length of the sarcomere. The explanation for this appearance is shown in Figure 9.

Each sarcomere contains two sorts of protein filament. There are thin filaments of **actin** and thick filaments of **myosin**. The different bands and zones are rather unimaginatively given the letters shown in Figure 9. (It is

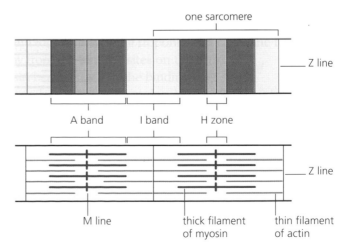

▲ **Figure 9** The arrangement of actin and myosin filaments in two sarcomeres.

Chapter 9
Chemicals in control

Hormones, and indeed chemicals in general, often get a bad press. Hormones are blamed for erratic behaviour in teenagers and for a variety of eccentricities in men and women. Hormones are frequently thought to be minor players in the body's communication system compared with nerves. As you have seen in the previous chapters, the nervous system depends on chemicals, for example at synapses and neuromuscular junctions. Many of the neurotransmitters are now known to be chemically very similar to hormones.

It was only in the 20th century that biologists discovered that chemical 'messengers' are transported round the body in the blood and that these play an essential part in the body's communication system. The term 'hormone' was not coined until 1906. About 50 years before this, a German physiologist called Arnold Berthold had the idea of finding out what happens when a testis is put back into the abdominal cavity of a castrated chicken. Farmers had been castrating male chickens for centuries because they fattened up better. People were well aware that a castrated cockerel failed to develop the large red comb on its head, but no one seems to have tried to explain why. Berthold discovered that the castrated chickens with an implanted testis developed a red comb like a normal cockerel. He deduced that the testis must have produced some substance that allowed the comb to develop, but he didn't explore this idea further. It was many years before sex hormones were discovered.

▲ **Figure 1** The development of a cockerel's comb is controlled by sex hormones produced in the testes.

Of course, it was well known that the testes were important for the development of male characteristics. For thousands of years Chinese emperors had been castrating young men who could then safely be trusted to look after their harems. In Europe the custom of castrating boys to preserve their singing voices was still being carried out in Berthold's time. Castrated boys fail to develop secondary sexual characteristics, such as a beard, pubic hair and a deeper voice at puberty.

Puberty is the time when a young person becomes sexually mature and capable of reproduction. Compared with other mammals this is remarkably late in humans – on average at about 12 years old in females and 14 in males. Maturation does not happen suddenly. It spans a period of several years, usually beginning somewhere between the ages 7 and 10, although at first the changes are not obvious. The male hormone testosterone and the female hormone oestrogen play important roles in the changes. We now know that many other hormones are also involved and that these originate in several different sites in the body. Perhaps surprisingly, males and females make both testosterone and oestrogen, although in different quantities and for different functions. The activity of the sex hormones is regulated by two different hormones produced in the pituitary gland, which is situated just below the brain. The same two hormones stimulate the ovaries to develop and release eggs in females and act with testosterone in the testes to produce sperms in males. The release of these hormones from the pituitary is controlled by other hormones that are secreted from a neighbouring part of the brain, the hypothalamus.

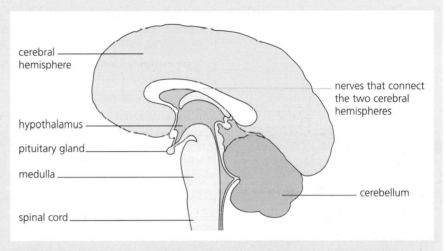

▲ **Figure 2** This shows a section through the centre of the brain, separating the two cerebral hemispheres. The pituitary gland and the hypothalamus have important roles in several different control systems. We shall be referring to them several times in this chapter so it will be useful to know where in the brain they are situated.

Maturation depends on a complex set of interactions between hormones. It is not surprising that things sometimes go awry. However, the suggestion that excess amounts of hormones can cause extreme mood swings or violent behaviour, especially in adolescent males, seems to be unjustified. For example blood tests show no correlation between testosterone concentration and aggressive behaviour. Recent research has shown that as well as the visible bodily changes that occur during adolescence there is also significant restructuring within certain parts of the brain. For example the neural networks that form in the frontal lobes of the brain undergo considerable changes. It is this part of the brain that you may remember was involved in the mood changes experienced by Phineas Gage (page 151). Perhaps the maturation of the brain is responsible for some of the erratic behaviour of which teenagers are often accused. We should be grateful that we are not butterflies! They undergo an almost complete breakdown of the organs of the caterpillar before being transformed into the reproductive adult phase of their life cycle. Hormones are responsible for these changes too.

In this chapter we shall look at some examples of the ways in which chemicals control physiological processes in humans and other mammals. One important role of hormones is maintaining more or less stable conditions in the body. Hormones ensure that the glucose concentration in the blood doesn't rise too high or fall too low. They keep the pH and water content of the blood fairly constant. Hormones also control the menstrual cycle. We shall also look at the action of some other chemicals that have only localised effects and so are not usually defined as hormones. The tropic responses of plants depend on chemicals, so we will explain how these operate.

What is a hormone?

A hormone is a substance that is secreted directly into the blood from the tissue where it is made. The blood carries it to all parts of the body but the hormone only has an effect on certain **target organs** or **target cells**. Many hormones work in a similar way to neurotransmitters. They attach to **receptor molecules** on the plasma membrane of target cells. This sets off reactions that activate enzymes in the cells, as we shall see later. Some hormones that are soluble in lipids can pass through the plasma membrane into the cell. Here they may enter the nucleus and directly stimulate genes to produce enzymes.

Hormones are synthesised in **endocrine** tissues. Endocrine **glands** are organs whose only function is to produce hormones. These include the pituitary gland situated below the brain (Figure 2) and the adrenal glands, one at the top of each kidney. Other hormones are made by groups of cells in other organs, such as the cells that produce the sex hormones in the testes and ovaries, and the insulin-making cells in the pancreas.

Q1

Not all glands are endocrine glands. Suggest why the salivary glands are not described as endocrine.

How are hormonal systems different to nervous systems?

Both hormonal and nervous systems depend on chemicals to produce their effect on the target cells and organs, but these chemicals are utilised in different ways.

- In the nervous system the chemicals are the neurotransmitters that only have to travel the few nanometres across a synapse or neuromuscular junction. Neurotransmitters are only released when impulses pass along the fibres of neurones to particular targets.
- Hormones are made in specialised endocrine cells and reach their targets via the bloodstream. Hormones can be synthesised and then stored for rapid release.

As a result, compared with hormones, nervous communication in general:

- is faster
- targets specific parts, such as a single muscle
- has its effect for a shorter time.

This does not mean that hormonal communication is particularly slow. For example, if you get a sudden fright adrenaline is released from its store in the adrenal glands. You soon feel a range of effects, such as a thumping heart and faster breathing. The adrenaline spreads to all parts of the body in the blood. This has the advantage that the hormone can have different effects on different targets. As well as stimulating the sinoatrial node and the intercostal muscles, it narrows the arterioles that supply blood to the skin and intestines and redirects the blood to the muscles. It also stimulates the breakdown of glycogen to glucose (page 207).

Some of the same responses, such as speeding up the heartbeat by stimulating the SAN, are produced by the **sympathetic** nerves. These are part of the autonomic system that operates our automatic and unconscious responses. As we saw in Chapter 7 (page 168) the synapses of these nerves produce a neurotransmitter that is chemically almost identical to adrenaline. The sympathetic nerves can increase the heart rate without the other effects of adrenaline, for example during exercise. Adrenaline has the advantage that it quickly produces a wide range of coordinated responses that prepare the body for emergency action. These effects can be maintained for as long as is necessary.

Histamine

If you suffer from hay fever you will be very likely to have used antihistamine drugs. Not surprisingly, these work against histamine. Histamine is produced by mast cells (Figure 3). **Mast cells** are quite large cells that occur in connective tissue.

Mast cells respond to tissue damage or infection by bacteria by secreting **histamine**. As you may remember from the AS course, this causes the tissue to become inflamed. The histamine makes the blood capillaries more permeable, so phagocytes can more easily pass through into the tissue fluid and attack the invading bacteria. Histamine has its effect close to the mast cells that produce it. It is not distributed through the blood system so it is not referred to as a hormone. Substances that cause localised responses are called **chemical mediators**.

Inflammation is an important part of the **immune response** and the rapid reaction prevents the infection of small scratches and injuries spreading throughout the body. However, it is not so helpful in people who are sensitised to particular allergens, such as dust or pollen. For them it can cause widespread irritation of the nasal passages and airways to the lungs. Another effect of histamine is that it causes the contraction of muscle cells in such places as the lining of the airways. This may have the natural advantage of preventing infective organisms penetrating deeply into the lungs, but it is a serious problem for asthmatics.

Prostaglandins

Like histamine, **prostaglandins** are chemical mediators. They too are secreted when a tissue is damaged. Almost all tissues of the body can produce them. Over a dozen different prostaglandins have been discovered, with a wide range of functions. As well as their role in the inflammatory response to tissue damage, they are also involved in blood clotting, the

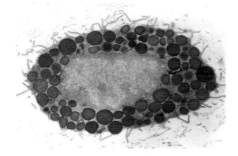

▲ **Figure 3** The cytoplasm of a mast cell contains many granules that secrete histamine. Mast cells can change shape and move short distances to areas of damaged or infected tissue. Magnification ×4000.

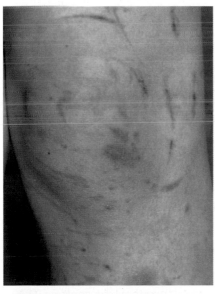

▲ **Figure 4** Inflammation is a first line of defence when the skin is damaged. Redness is due to increased blood flow in the skin capillaries. The surrounding area swells because fluid leaks from the capillaries as they become more permeable due to the effect of histamine. Phagocytes migrate from the capillaries into the tissue and destroy bacteria that may have entered the skin through the wound. Dead phagocytes and bacteria form pus.

Q2

Explain how histamine secretion may cause an asthma attack.

control of blood pressure and the production of secretions in the digestive system. One prostaglandin is contained in human semen. It makes the muscles in the walls of the uterus and oviduct contract during female orgasm, thus speeding the movement of the sperms towards the ovum. Exactly how the prostaglandins work is not yet fully understood, but some of them act in association with a hormone inside their target cells.

Control of growth

Charles Darwin had an insatiable curiosity. In his old age, years after writing his famous book on natural selection, he turned his attention to the mysteries of growth in plants. One thing that intrigued Darwin was how shoots always grow up and roots grow down, no matter which way up a seed is planted. He experimented by germinating seeds that he fixed upside down. He and his son, Frank, watched them for hours and traced the movement of the growing root tip. Gradually the root bent round and started to grow down in a series of tiny zigzags. He thought the zigzags might be due to vibrations. To test this hypothesis he got his son to play his bassoon at them. This made no difference to the growth pattern, so Charles dismissed the suggestion. He was forced to conclude that there was some inbuilt driving force but he had no means of investigating the chemistry.

Investigating growth and tropisms

Some 50 years later, after hormones had been identified in animals, researchers started to look for similar chemical influences in plants. Figure 5 summarises the results from a range of different experiments. These were carried out on the shoots of cereal plants, such as maize or oats. These shoots (called coleoptiles) have the advantage that they grow straight up with the young leaf concealed inside. The actual growing region is a few millimetres below the tip.

Study the results from five sets of experiments in Figure 5. Answer the questions about each set before referring to the explanations.

■ Investigation 1 shows the results of an experiment in which the tips of shoots were cut off just above the growing region. The tip was put back on to the cut end of half of the decapitated shoots. Describe the results and explain what the results show.

When the tip was not replaced the shoots stopped growing. The shoots with the tips replaced did grow, but more slowly than the controls. This suggests that the tip is responsible for stimulating growth, since its removal stops further growth. Some signal must pass from the tip to the growing region. This could be along some pathway similar to a nerve or by means of a chemical that passes down the shoot. Replacing the tip restores the ability to grow, so it is more likely that it is a chemical that passes down to the growing region. The cut may reduce the rate at which the chemical moves.

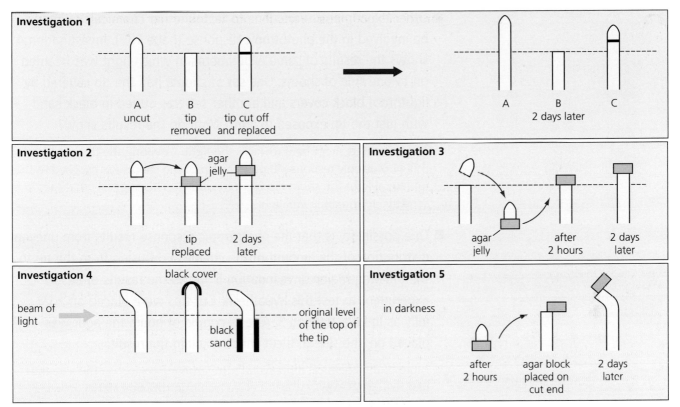

▲ **Figure 5** Experimental results from five investigations into growth and tropic responses in wheat coleoptiles.

■ Investigation 2 shows the results of a follow-up experiment. The tips were cut off and replaced, but this time a thin slice of agar jelly was placed between the tip and the cut shoot. The shoots were still able to grow. What does this show about the pathway?

The signal must be able to pass through the agar jelly. A possible explanation is that a chemical is diffusing through.

■ Investigation 3 shows the results of a further refinement. After cutting off the tip it was placed on a small piece of agar jelly for a couple of hours. The agar jelly was then put on the cut end of the shoot without the tip. What can you conclude from the results?

A substance must have passed into the agar from the tip. This substance must have diffused into the cut end of the shoot from the agar.

■ Suggest a control that should be carried out in the experiment shown in Investigation 3.

On the cut end of a shoot place a block of agar jelly that has not had a cut tip on it. This will confirm that there is not something in the agar that stimulates growth.

■ Other experiments were done to test whether chemicals might also be involved in the phototropic response (page 139). Investigation 4 shows the results of some experiments in which light was beamed on to one side of shoots. One set of shoots had the tip covered by lightproof black covers and another set was buried in black sand with just the tip exposed to light. What do the results show?

The untreated shoot bent towards the light. Covering the tip prevented this phototropic response. If only the tip of the shoot was exposed to the unilateral light the shoot did bend towards it. This suggests that the tip is sensitive to the unilateral light.

■ One possibility is that the phototropic response results from uneven distribution of the amount of the chemical diffusing from the tip to the growing region. Investigation 5 shows the results of an experiment to test this hypothesis. Cut tips were placed on agar jelly as in Investigation 3. After a couple of hours the agar was placed on one side of a cut shoot. Explain the result.

More of the chemical diffused into the growing region on the side that had the agar. Growth was therefore faster on this side so the growing shoot bent over.

■ Bearing in mind the information above, suggest two hypotheses that could explain how unilateral light causes an uneven distribution of the chemical and produces the phototropic response.

One possibility is that the chemical is destroyed or partially destroyed on the illuminated side, so more passes down to the growing region on the shaded side. Another is that shade stimulates the production of the chemical. It is also possible that the chemical diffuses away from the illuminated side and the concentration on the shaded side is increased in this way. Finally, it is possible that in bright light the action of the chemical is affected by an inhibitor. We will come back to these ideas later.

Plant growth regulators (PGRs)

Most of the research described in the previous section was carried out in the 1920s. The next step was to identify the chemical involved in the growth response. This proved to be extremely difficult. The first substance to be discovered was **indoleacetic acid (IAA)**. Several other substances that affect growth in plants have been discovered since. But there is still uncertainty and disagreement about exactly how these substances work. One of the problems is that the actual concentrations of the substances present in the plant tissues are extremely low.

Growth of a root or shoot has two distinct stages.

- First, cell division takes place at or near the tip.
- Second, the cells elongate. This starts in the cells a short distance behind

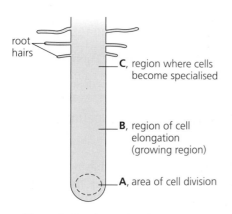

root hairs

C, region where cells become specialised

B, region of cell elongation (growing region)

A, area of cell division

▲ **Figure 6** Growing regions in a root.

What type of cell division would occur in the tip (region **A** in Figure 6)?

Q4

In region **C** cells become specialised for particular functions. What is this process called and what tissues will be produced?

Q5

When the substances that control growth in plants were first discovered they were called plant hormones. This name has been abandoned and the substances are now called plant growth regulators (PGRs). In what ways are PGRs **a)** similar to hormones in animals, **b)** different?

the tip in the area that we have been calling the growing region in the earlier descriptions. This is where the most obvious increase in length of a root or shoot occurs.

IAA has its main effect in the elongating region. It is synthesised in the young root cells at the tip. As it diffuses into the elongating region it attaches to receptors on the membranes of cells. Exactly how it works is still unclear, but one of its effects is to lower the pH by the release of hydrogen ions. These break some of the bonds between the microfibrils in the cellulose walls (page 140 of the AS book), making them more easily stretched by the turgor of the cells. There is also evidence that IAA can increase the rate of transcription in the cells and thus speed up protein synthesis. However, adding IAA artificially to shoots speeds up elongation within 15 minutes, so it seems unlikely that an increase in protein production could have a major effect so rapidly.

IAA and tropisms

On page 192 we suggested some possible explanations for the phototropic response in plant shoot tips. In summary these include:

- IAA is destroyed by the light on the illuminated side
- extra IAA is produced on the shaded side
- IAA moves away from the illuminated side
- the action of the IAA is inhibited on the illuminated side.

Figure 7 shows results from some more recent experiments to investigate these ideas. In each case the shoot tips were cut off and then placed on thin blocks of agar. In some experiments very thin slices of mica were used to separate the two sides of the tips and/or the agar blocks. After 3 hours the total amount of IAA in the agar blocks was measured.

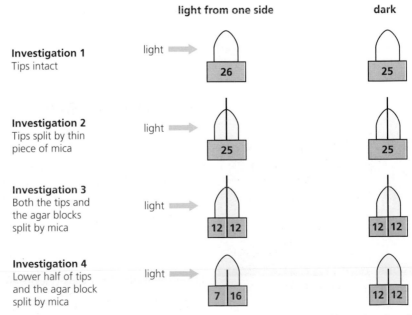

▲ **Figure 7** Results of four experiments on shoot tips to investigate possible explanations for their phototropic response. The figures show the amounts of IAA in arbitrary units.

■ **Against which of the explanations does Investigation 1 (Figure 7) give evidence?**

The amount of IAA in the agar blocks is about the same whether in the light or dark. If it is destroyed on the illuminated side you might expect an overall decrease in the light. If more was produced on the shaded side you might expect an increase in the dark.

■ **What do Investigations 2 and 3 show?**

They confirm that the total amounts of IAA passing into the agar are the same in both dark and light conditions and that the amounts passing down the illuminated and shaded sides are the same.

■ **What does Investigation 4 show?**

In darkness the amounts passing down each side were the same. In unilateral illumination more IAA passed down the shaded side. This suggests that when the tip is illuminated the IAA produced does move towards the shaded side. You will notice that the total amounts passing down are the same as in the other experiments. It would therefore seem unlikely that the IAA is inhibited on the illuminated side.

Further research has shown that, as so often in living organisms, the process is much more complicated than was originally thought. Experiments with radioactively labelled IAA have shown that it is moved across a tip by carrier substances. Several other factors are also involved, including the wavelength of the light and other plant growth regulators (PGRs). This is a good example of an obvious explanation eventually proving to be incorrect when thoroughly investigated.

Making use of PGRs

Selective weedkillers

Soon after IAA was discovered scientists realised that it could have commercial uses. The amounts produced by plants were far too small for useful quantities to be extracted. Chemists set about making IAA, or chemically similar compounds, that would have the same biological effects. Ideally it would be preferable to synthesise the natural compound, but in practice this often proves to be rather unstable. Two types of artificial products were developed. One has been a great success as a constituent of rooting powders. These are used to encourage plant cuttings to develop roots and are of great importance in the horticultural industry. Other products have a wide use as weedkillers. They work on the principle that large doses of the growth regulator disrupt growth so much that the plants grow abnormally and die. Broad-leaved plants, which most weeds are, absorb larger quantities of the weedkiller than narrow-leaved plants such as grass and cereal crops (Figure 8). This makes them very useful as **selective weedkillers**. They can be used to kill weeds in lawns and cereals without affecting the grass or crop, provided appropriate amounts are applied.

▲ **Figure 8** Distorted growth in a broad-leaved weed in a wheat field resulting from selective weedkiller application. Note the wheat plants are unharmed.

▲ **Figure 9** These two monster cabbages were treated with gibberellins. You can see that the stems have grown abnormally long, so the leaves are far apart compared with those of the normal cabbages beside them.

Stimulation of fruit formation

Another function of IAA in plants is to stimulate the formation of fruits after fertilisation. Spraying the flowers of apple and pear trees with compounds similar to IAA allows fruit growers to make sure that the fruits develop even in the absence of pollinating insects. Other PGRs have major commercial uses. **Gibberellins** were discovered to have a significant effect on the growth of stems, as you can see in Figure 9.

This feature is not particularly useful in growing plants, since few people want 12-foot-tall cabbages! However, the brewing industry uses gibberellins to stimulate the germination of seeds. Brewers produce the malt sugars used for fermentation by growing barley grains for a few days. The starch is broken down to sugars more quickly when gibberellins are added. Gibberellins are also used to produce the seedless grapes that are much favoured by consumers.

Homeostasis

We have just seen that plants do have complex control systems. These are essential for their survival, as we mentioned in Chapter 6. The phototropic response ensures that plant stems and leaves are orientated towards the light to maximise the rate of photosynthesis. Plants can survive in a huge range of environmental conditions. In winter they can survive in extremely low temperatures. They may not grow much in the lowest temperatures, but they certainly can grow in temperatures that are much lower than would be comfortable for our bodies. It is a myth that enzymes only work at about 37 °C. In plants they operate at much lower temperatures than this and their optimums are often also much lower.

Animals, of course, are much more active. Most have to move around to obtain food. For the majority the environmental temperature determines their ability to carry out their activities. Even then many are capable of activity at remarkably low temperatures. There are, for example, fish that live in deep seas where the temperature is constantly only a few degrees above freezing.

Humans, however, and other mammals and birds are adapted to maintain high levels of activity whatever the environmental temperature. Our **core body temperature** stays more or less constant at approximately 37 °C. In birds it stays rather higher than this, somewhere between 40 and 42 °C. The benefit is that mammals and birds can feed and grow rapidly at all times. There is, however, a price to pay. They use more energy and need more food. They therefore require reliable sources that can provide them with enough food at all times. Nevertheless, in evolutionary terms, the ability to maintain a high temperature has given them significant advantages.

The ability to keep conditions inside the body more or less constant is called **homeostasis**, which roughly means 'standing still' in ancient Greek. If the temperature falls mammals and birds maintain their internal body temperature by generating heat from respiration. For this reason they are said to be **endothermic**, that is 'producing heat inside'. Other animals have to gain heat from their environments and they are called **ectothermic**.

In addition to the internal temperature mammals have also to keep the following constant.

* Water potential of the blood and tissue fluid: this is kept constant by regulating the water content and the concentrations of glucose, ions and other solutes. A rise in the concentration of solutes will cause problems with osmosis. Tissues will lose water into the more concentrated blood plasma and become dehydrated.
* pH: this must be maintained within narrow limits. An increase in hydrogen ion concentration (i.e. a drop in pH value) can interfere with enzyme action by disrupting their protein structure (see the AS book, page 23).

We will look at some of these examples of homeostasis later, but first we will see how the temperature is maintained.

Temperature regulation

Stand outside with no clothes on when the temperature is around freezing point and you will immediately feel cold because temperature receptors in your skin are stimulated. For a time your core body temperature will not change, although you will be unaware of this. However, you will rapidly lose heat from your body surface. The blood supply to the skin surface will be cut off and you will start to shiver. The body will release more heat internally and thus maintain your core body temperature. It may remain at about 37 °C for half an hour or more, but after a while you will be unable to generate enough heat to prevent a decline. Hypothermia sets in. When the core temperature falls to about 32 °C most people lose consciousness. Death occurs in most cases at about 25 °C. The rate of cooling is much greater in cold water, so if you are unlucky enough to fall into the Arctic Sea, death is likely within 15 minutes, even if you don't drown first. Compared with most mammals, humans are very poorly adapted to withstand low temperatures, mainly because of the absence of hair. Some mammals, such as the Arctic fox, can maintain their core temperature without having to produce additional heat internally, even when the temperature is as low as −40 °C.

In the UK, we have to maintain our body temperature well above that of the external environment. In averagely comfortable conditions the temperature is likely to be only about 20 °C, so when a person is naked there is a constant loss of heat energy to the environment. Even in the tropics, where humans first evolved, the atmospheric temperature is usually below that of the body. In our climate we have adapted our behaviour by the use of clothing, buildings and external sources of heat.

Negative feedback

To keep the core body temperature constant it must be possible to detect changes in blood temperature.

* If the temperature falls: extra heat must be generated and heat loss must be reduced.
* If the temperature rises: more heat must be lost from the body surface. Figure 11 shows this diagrammatically.

▲ **Figure 10** Arctic foxes can maintain their core temperature without having to produce additional heat internally, even when the temperature is as low as −40 °C.

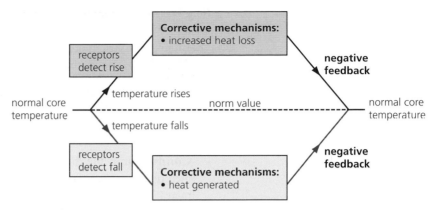

▲ **Figure 11** Negative feedback plays an important role in thermoregulation in endotherms.

The processes shown in Figure 11 illustrate an important principle that we shall see in other examples of homeostasis. This is **negative feedback**. In any negative feedback system a change is detected and then effectors return the system to its original state. Central heating systems operate in the same way. A thermostat detects a drop in temperature in a room. This switches on the heating. Once the room temperature is back to the pre-set level the thermostat switches the heating off.

How heat is generated in the body

Respiration, as you will recall from Chapter 2, page 41, is nowhere near 100% efficient. The energy released from respiration in muscles is not all used for contraction of the fibres. Much is released as heat. Therefore any form of muscular activity will generate heat, as will all the processes that use energy from respiration, such as enzyme synthesis and active transport. Even when the body is totally at rest, the **basal metabolic rate** will produce heat, for example from the heartbeat, the muscular activity of breathing and the manufacture of proteins for growth and enzymes. During very vigorous exercise the core temperature may go up by 2 or 3 degrees before cooling mechanisms can return it to the norm.

If the body's core temperature does fall, heat can be produced by increasing the metabolic rate. Release of the hormone **adrenaline** from the adrenal glands produces an increase in the metabolic rate throughout the body. This will include respiring some of the stores of glycogen in the liver.

Most mammals have some stores of **brown fat** that are able to release energy very rapidly. In humans brown fat deposits are scattered throughout the body but with a concentration in the upper part of the back between the shoulder blades. Unlike the much larger stores of white fat used as insulation below the skin and round many organs, brown fat has sympathetic nerve connections that can stimulate its breakdown to fatty acids. Brown fat contains many mitochondria which can respire the fatty acids for rapid heat production.

The body temperature can also be raised by **shivering**. This is the result of small and rapid muscular contractions that are automatically activated by the brain when the blood temperature falls. The extra muscular activity is another way of generating heat.

How heat is lost from the body

There are four processes by which heat can be lost from the body: conduction, convection, radiation and as a consequence of evaporation.

- **Conduction** is the direct transfer of heat energy to cooler materials, including air, water or solid objects, such as stone or metal. How much heat energy is lost in this way depends on how well the material conducts the heat away from the body surface. Air is a poor conductor, so the air near the skin heats up only slowly. Provided the warm air is not moved away, air acts as a good insulator. Water is a better conductor so body heat is lost more rapidly into water. Metals quickly conduct heat away. They feel cold when we touch them because heat is conducted away from the skin rapidly, even though the metal is at the same temperature as the air around it.
- **Convection** is the movement of currents of warm air. Because warm air is less dense than cooler air it tends to rise. Although air itself is a poor conductor the layer of warm air that is created on the outside of the skin rapidly rises away unless trapped by fur or clothing.
- **Radiation** is the transfer of heat energy by infra-red waves. This is what you experience when you feel the heat at some distance from a fire or a radiator. It is the process by which we lose the majority of heat from the body, usually 50 to 60% at room temperature. It is also the way in which we can gain heat if there is a hotter environmental source, such as a fire or sunshine.
- **Evaporation** is the conversion of liquid water to vapour. This process requires a lot of heat energy. Water evaporating from the skin's surface gains this heat energy from the skin. This can provide a significant cooling effect; about a quarter of heat is lost in this way in the right conditions.

The body's cooling systems

Control of our body temperature depends on adjusting the gain or loss of heat. If the core temperature starts to rise we speed up the loss of heat from the skin. Figure 13 shows the structural adaptations of the skin that make this possible.

- **Vasodilation** As you can see from Figure 13 the skin has a network of capillaries just below the protective outer epidermis. The capillaries are supplied by arterioles situated lower down in the dermis. Rings of muscle, called **sphincters**, in the arterioles can control the blood supply to the capillaries. When the body temperature rises the sphincters open allowing blood to flow through the capillaries near the surface. This is called vasodilation (Figure 14). The process allows more heat to be lost by radiation, and with a warmer surface conduction and convection from the skin will also increase.
- **Vasoconstriction** When the blood temperature falls below the norm the sphincters close and the flow of blood through the capillaries is slowed

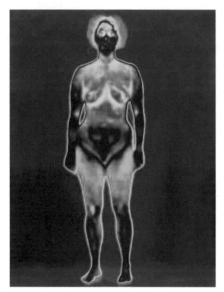

▲ **Figure 12** Thermal imaging shows the main areas of heat loss from the body. The yellow areas are losing the most heat. The green/blue halo around the head shows heat loss by convection.

Q_6

It is incorrect to say that the capillaries constrict. Explain why.

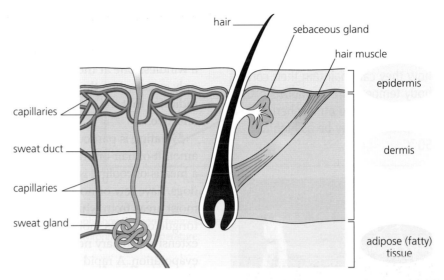

▲ **Figure 13** How the skin is adapted for speeding up or slowing down heat loss.

Q7

Prolonged exposure to low temperatures can result in damage and death of skin cells. Suggest why.

or stopped. This is called vasoconstriction (Figure 14). In some exposed areas of skin, such as in the fingers and ears, there are **shunt vessels** that bypass the capillary network by connecting the arterioles to venules. These also can be opened and closed by circular muscles. (Note that it is quite wrong to suggest that capillaries move up and down in the skin. Blood either flows through the capillaries when we need to lose heat or is prevented from flowing through to preserve heat. It would be impossible for capillaries to push their way to or from the surface of the skin.)

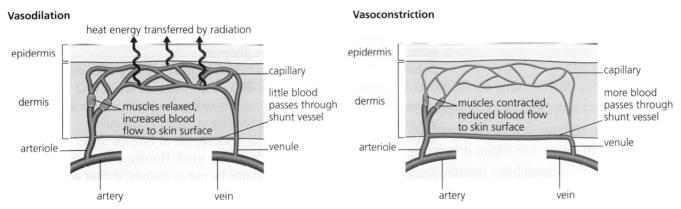

▲ **Figure 14** Vasodilation and vasoconstriction in capillaries in the skin.

- **Sweating** In high temperatures the **sweat glands** secrete a slightly salty solution on to the surface of the skin. There are several million sweat glands distributed over all parts of the skin. Sweat does not just leak out through pores that open or close, as is often thought. The sweat glands contain cells that actively pump the sweat into the coiled tube. This process is stimulated by branches of sympathetic nerves. As the water evaporates it cools the skin surface and the blood flowing through the surface capillaries. Sweating is only effective when the sweat can

▲ **Figure 18** This anaconda has just made a meal of a caiman (a type of alligator), which it has swallowed whole. The anaconda may not need to feed again for several months.

muscles generate enough heat to keep the temperature above that of the surroundings. They also have physiological adaptations. For example the optimum temperature for their enzymes is usually lower than that of endotherms and often is adapted to suit the average temperature in the climate where they live.

So why don't ectotherms evolve into endotherms? Ectotherms have several advantages. They need much less food than endotherms because they don't use large amounts to generate heat. Small mammals, such as shrews, need to eat almost constantly because their large surface area/volume ratio makes it very difficult for them to retain heat. Large reptiles, such as crocodiles and big snakes (Figure 18), can manage on a single large meal for weeks. This makes them much less vulnerable to fluctuations in the food supply.

How does the temperature control system work in endotherms?

The temperature of the blood is monitored by the **hypothalamus** in the brain, which we mentioned in the chapter introduction (Figure 2, page 187). Receptors in the hypothalamus detect a rise in temperature of the blood flowing through it. They transmit impulses to a part of the hypothalamus called the **heat loss centre**. This centre then sends impulses to the sphincter muscles in the skin arterioles and to the sweat glands. Vasodilation and increased sweating result. The blood is cooled and the core temperature falls.

Conversely, if the blood temperature falls, receptors stimulate the **heat gain centre** in the hypothalamus. This stimulates vasoconstriction and shivering. The hairs on the skin will be pulled upright (although this is of little importance in humans). The hypothalamus is therefore the 'thermostat' that controls body temperature. It is cleverer than a room thermostat in that it doesn't just switch on the heating system when the temperature falls but also switches on the air-conditioning when the temperature rises.

We can now add more detail to the diagram of the negative feedback system (Figure 11, page 197).

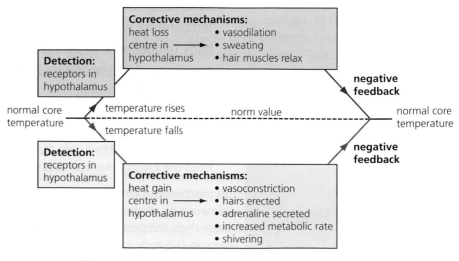

▲ **Figure 19** The hypothalamus controls the negative feedback system in thermoregulation in endotherms.

Positive feedback

The graph in Figure 20 shows the effect of environmental temperature on the metabolic rate of a naked human. The measurements were made while the person was lying down at rest. A value of 100 units on the scale is the basal metabolic rate.

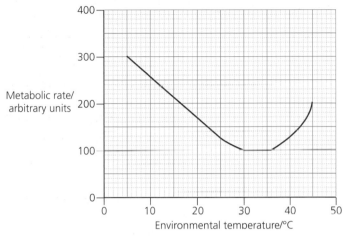

▲ **Figure 20** The effect of temperature on metabolic rate in a naked human.

■ **The metabolic rate stayed constant when the environmental temperature was between about 29 and 36 °C. Explain why.**

The environmental temperature is only a few degrees below the body temperature. The rate of heat loss is very low. The insulating layer of fat under the skin and vasoconstriction can maintain the body temperature. No extra energy needs to be generated at the lower end of the temperature range. The cooling processes, such as vasodilation and sweating, can keep the body temperature constant at the higher end.

■ **At 16 °C the metabolic rate was roughly double the basal rate. Explain why the rate was much higher than at 29 °C.**

The difference between the skin temperature and the environment was too great for insulation and vasoconstriction to prevent heat loss. Therefore heat energy was generated from respiration to maintain the core body temperature. This may have involved shivering, an increased rate of respiration in the liver and respiration of brown fat stores.

■ **What happens to the metabolic rate when the environmental temperature goes above 37 °C?**

The metabolic rate increases and gradually the rate of increase goes up.

■ **Suggest an explanation for the increase in the rate of metabolism.**

The body temperature is not fully controlled by the normal cooling mechanisms. As the environmental temperature becomes greater than the body temperature heat cannot be lost by radiation, convection or conduction. Only evaporation of sweat can keep the temperature down.

■ **Suggest a factor that would affect how rapidly the metabolic rate would rise.**

The humidity of the air would affect the rate of evaporation and its effectiveness in cooling.

■ **Suggest why the rate of increase gradually goes up faster.**

As the core temperature rises the rate of respiration also rises, because the rate of chemical reactions increases as the temperature goes up. This has the effect of releasing even more heat energy, so the temperature is raised even more. As the temperature continues to rise, the rate of increase gets even faster. This is an example of **positive feedback**.

A rise in core temperature above about 42 °C is a life-threatening condition. This is called heat stroke. It needs emergency treatment to bring the temperature back down. The activity of cells, especially in the brain, is seriously disrupted. This is usually due to a combination of factors. Commonly it is the effect of dehydration and ionic imbalance in the blood. Although there may be some increased denaturation of enzymes, this is unlikely to be a major factor at this temperature.

■ **Positive feedback may also play a part in making worse the effect of a fall in core temperature. Suggest how.**

As the temperature falls the rate of respiration decreases. Therefore less heat can be generated by the respiration in the liver and elsewhere so it becomes more difficult to maintain the body temperature.

Control of blood glucose concentration

The control of the concentration of glucose in the blood is another example of negative feedback in homeostasis. The set value for the concentration in the blood is approximately 90 mg per 100 cm^3. It is important that there is always some glucose circulating in the blood because the cells need a constant supply for respiration. The brain cells in particular will very soon die if their supply is cut off. On the other hand, if the concentration of glucose rises too high it has a major effect on the water potential of the blood.

Glucose enters the blood from three main sources:

* absorption from the gut following the digestion of carbohydrates
* breakdown of stored glycogen
* conversion of non-carbohydrates such as lactate, fats and amino acids.

The amount of glucose absorbed into the blood from digestion can fluctuate massively. The control system therefore has to get rid of the excess glucose entering the blood after a carbohydrate-heavy meal and to release

Q10

Explain the effect of high glucose concentration on the water potential of the blood.

glucose rapidly from storage compounds when muscles are depleting the content of the blood at a fast rate during exercise. The negative feedback system operates on the same principle as in temperature regulation, except that this time hormones make the correction instead of nerves.

What happens when blood glucose concentration rises?

A rise in the blood glucose concentration above normal is detected by the **beta cells (β-cells)** in the **pancreas**. The β-cells are situated in little groups of cells dotted around the pancreas called **islets of Langerhans** (Figure 21). The β-cells synthesise a hormone called **insulin**. You will probably recall that the pancreas secretes digestive enzymes that pour through a duct into the upper part of the small intestine. Since they produce hormones and not enzymes, the islets are like mini-endocrine glands.

When blood with a high concentration of glucose reaches the islets, glucose is absorbed into the β-cells. The plasma membrane of a β-cell contains carrier protein molecules that transport glucose into the cells by facilitated diffusion. This stimulates vesicles of insulin to move to the membrane and release insulin into the capillaries. This is similar to the process at synapses where acetylcholine is released into the synaptic cleft.

Q11

What is meant by facilitated diffusion? (If you can't remember you may need to refer back to your AS course.)

Q12

How does facilitated diffusion differ from active transport?

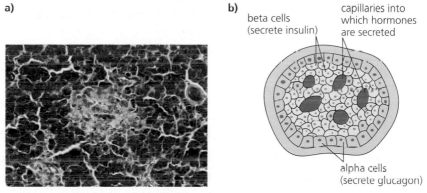

▲ **Figure 21 a)** Coloured photomicrograph of islets of Langerhans in the pancreas, magnification ×150; **b)** diagram of an islet of Langerhans.

Insulin then circulates round the body in the bloodstream. It stimulates uptake of glucose by cells in muscles, adipose (fat storage) tissue and the liver. Insulin attaches to receptor molecules in the membranes of the cells in these tissues. Glucose cannot diffuse into cells through the phospholipid bilayer. All cells have carrier proteins that allow glucose to enter by facilitated diffusion, but the rate of uptake is limited by the number of carrier molecules. In muscles and adipose tissue insulin causes additional carrier proteins to join the membrane (Figure 22, overleaf). By adding many more carrier proteins the rate of uptake of glucose from the blood is greatly increased.

Liver cells already have large numbers of carrier proteins in their membranes. In the liver insulin speeds up glucose uptake in a different way. After the glucose has entered the liver cells it activates an enzyme that rapidly converts the glucose to glucose phosphate. This maintains a steep **diffusion gradient** between the blood and the liver cells. Other enzymes

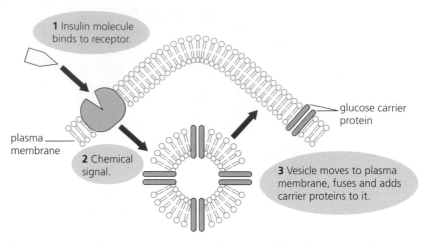

1 Insulin molecule binds to receptor.

plasma membrane

2 Chemical signal.

glucose carrier protein

3 Vesicle moves to plasma membrane, fuses and adds carrier proteins to it.

▲ **Figure 22** In muscles and adipose tissue insulin causes additional glucose carrier proteins to join the plasma membrane.

then synthesise the glucose phosphate to glycogen, a process known as **glycogenesis**. Glycogenesis also occurs in the muscles, replenishing the stores there. In fat storage tissue insulin activates enzymes that manufacture fatty acids and glycerol, which are then stored as fat.

To summarise, insulin removes excess glucose from the blood by:

* increasing the rate of facilitated diffusion into muscle and fat storage cells by adding carrier proteins to the cell membranes
* increasing the rates of glucose uptake and glycogen synthesis in the liver
* increasing the rate of fat synthesis in storage tissue.

What happens when blood glucose concentration falls?

When the glucose concentration in the blood drops below approximately 90 mg per 100 cm^3 insulin secretion ceases. The lower concentration stimulates **the alpha cells (α-cells)** in the islets of Langerhans to secrete another hormone, called **glucagon** (Figure 21). Glucagon's main effects are in the liver. It activates enzymes that break down the stored glycogen to glucose, which is then released into the blood. It can also activate enzymes that convert other substances to glucose, notably lactate and amino acids. This is a process called **gluconeogenesis**, which roughly means 'making glucose from new substances'. Both of these effects increase the blood glucose concentration again so that it quickly returns to its normal value. In practice, there are constant adjustments to the amounts of insulin and glucagon being secreted. Much of the time both hormones are secreted in small quantities with the proportions adjusted to maintain the glucose concentration at a fairly constant level.

Figure 23 summarises how negative feedback mechanisms operate in the control of the blood glucose concentration.

Q13
After its secretion, glucagon breaks down within a few minutes. Suggest the advantage of this.

Q14
The veins from the small intestine and the pancreas connect to the hepatic portal vein. This transports blood directly to the liver. (See diagram of blood circulation in the AS book, page 156.) Therefore the blood flows through the liver before returning to the heart and entering the general circulation. Suggest the advantages of this arrangement in relation to the control of the blood glucose concentration.

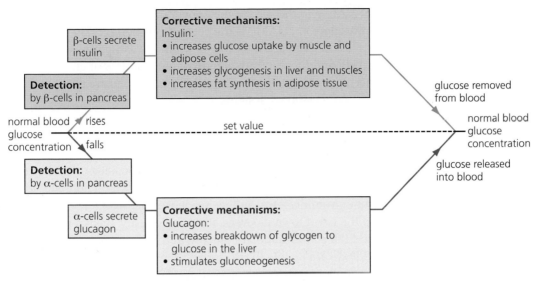

▲ **Figure 23** Negative feedback in the control of blood glucose concentration, showing the effects of insulin and glucagon.

The second messenger model

As often happens in scientific research, continuing investigations have revealed that the simple story does not give the full picture and that systems are more complex than at first thought. The breakdown of glycogen in the liver is not only stimulated by glucagon. Adrenaline, which, as we mentioned in Chapter 6, is the 'fight or flight' hormone, has the same effect. Adrenaline does not have a direct effect on the glycogen in the liver cells. As you can see from Figure 24 it attaches to receptors that span the plasma membrane. This activates an enzyme on the inside of the plasma membrane. This enzyme acts on ATP, which has three phosphate groups. Instead of producing ADP, the enzyme removes two of the phosphate groups. This makes **cyclic adenosine monophosphate**, usually written as **cAMP**. The cAMP then activates other enzymes that carry out the breakdown of glycogen to glucose. This process is called the **second messenger model**.

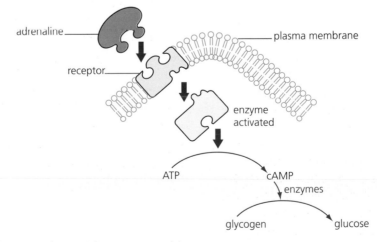

▲ **Figure 24** The second messenger model.

Q15

Suggest why it is an advantage for adrenaline to have its effect through a second messenger.

According to the second messenger model, the hormone is the first messenger and cAMP is the **second messenger**. The advantage of this system is that each molecule of hormone can produce many molecules of cAMP and these in turn activate large amounts of enzymes. Several other hormones, including glucagon but not insulin, are thought to use similar second messenger systems.

Diabetes

Diabetes is a condition in which the homeostatic control of the blood glucose concentration breaks down. Over two million people in the UK suffer from diabetes, and its incidence is growing. The increase is attributed to the trend towards over-eating and obesity. Insulin function is disrupted and the concentration of glucose in the blood rises. One effect is that the kidneys are unable to reabsorb all the glucose into the blood as they normally do. Therefore glucose appears in the urine. In response the kidneys tend to produce excess quantities of urine. This dehydrates the body and makes the sufferer very thirsty.

There are two main types of diabetes, called **Type I** and **Type II**.

Type I diabetes

Type I diabetes is caused by an inability to make enough insulin. It usually develops quite early in life, often in childhood. It is an **auto-immune condition** in which the immune system destroys some of the body's own cells. In the AS course we explained how T-cells attack infected cells by attaching to antigens on the plasma membrane. For some unexplained reason in Type I diabetes T-cells start to attack and destroy the β-cells in the islets of Langerhans. The shortage of insulin causes fatigue because not enough glycogen is stored in the liver. The concentration of glucose in the blood can rise to dangerously high levels after a meal. This can affect the brain and may result in coma and death.

This type of diabetes has to be treated by injections of insulin. Before the discovery of insulin, Type I diabetes was always fatal. For many years insulin was extracted from pancreas tissue of cattle and pigs. In the 1980s synthesis of human insulin by recombinant-DNA technology was one of the early successes of genetic engineering (Chapter 12, page 249). Treatment has been transformed by the development of fast-acting soluble and slow-release insulin preparations. These allow either for rapid control at a mealtime or control that can last for many hours. Methods of injection have been made much more convenient, for example by the use of injector pens. As a result people with Type I diabetes can live a normal and full life. Nevertheless most have to inject themselves with insulin twice a day in order to maintain a stable blood glucose concentration. There is hope that transplants of islets of Langerhans will provide more permanent treatment, but so far this has been much less successful than, for example, kidney transplantation.

Q16

Insulin is a protein. Explain why it has to be injected instead of being taken orally.

Type II diabetes

This form of diabetes is much more common than Type I; over 90% of cases in the UK are of this type. Usually it only develops in people over 40 years old but it is increasingly being found at younger ages. It is caused, not by failure to produce insulin, but by reduced sensitivity of the liver and fat storage tissue to insulin. The receptors either fail to respond or are reduced in numbers so glucose uptake is much reduced. The exact effects are still unclear because they are often associated with a range of other factors, often resulting from obesity and unbalanced diets. One response is that the β-cells are stimulated to produce larger quantities of insulin. If the condition persists the β-cells are damaged and the condition becomes more like Type I.

In the early stages Type II diabetes can be treated simply by careful attention to the diet. Reducing the intake of refined carbohydrates, such as sugars and forms of starch that are rapidly digested to sugars, avoids large surges in blood sugar concentration. The fat content of the diet also needs to be kept low and exercise helps to improve the body's sugar metabolism.

The glucose tolerance test

One of the tests used in diagnosing diabetes is the **glucose tolerance test**. After a period of several hours without food, the person being tested has a drink of water containing 75 g of glucose. The concentration of glucose in the blood plasma is measured at 30-minute intervals. The graph in Figure 25 shows the results of this test from three adults. One had no symptoms of diabetes. Both of the others had tested positive for glucose in the urine.

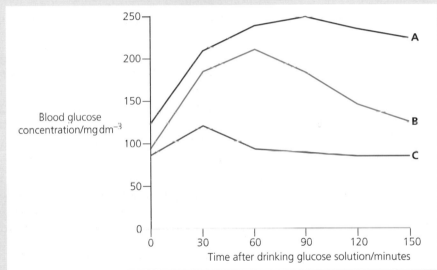

Figure 25 Results of a glucose tolerance test taken by three adults, labelled **A**, **B**, **C**.

1 Curve **C** shows the results for the non-diabetic person. Suggest why the glucose concentration rose in the first half-hour.

2 Describe and explain the results for curve **C** after 30 minutes.

3 Describe the differences between curves **A** and **B**.

4 Of the people **A** and **B**, one was diagnosed as having Type I diabetes and the other Type II. Which was which? Explain your reasoning.

The oestrous cycle

In the chapter introduction on page 187 we referred to some of the hormones involved in developing sexual maturity during puberty. In males hormones such as testosterone maintain a constant process of sperm production throughout much of adult life. In females the formation and release of ova takes place in a sequence of cycles controlled by a combination of hormones and involving both negative and positive feedback. The technical term for each cycle is the **oestrous cycle**, although in humans it is commonly known as the **menstrual cycle**, because each one lasts about a month. In other mammals the cycles can be quite different lengths. In a mouse, for example, the cycle lasts for four or five days. Some mammals have lengthy gaps between cycles, so females are only fertile for short periods at certain times. For example, red deer are fertile only once a year. This presumably is an adaptation that ensures that young are only born at favourable times of the year. Dogs normally have two cycles per year. It is during these times that the female is fertile and comes into **oestrus**, or 'heat' as it is commonly known. In fact it is from this behaviour that the term 'oestrous cycle' derives – it comes from the Greek word meaning 'frenzy'.

The menstrual cycle in humans

When a female baby is born her ovaries contain all her future egg cells. At puberty there are estimated to be about 300 000 cells that have the potential to mature into ova. During each menstrual cycle usually only one of the immature cells develops into a mature ovum. Occasionally there may be two, in which case twins may develop.

The essential function of the cycle is that it coordinates the release of an ovum from an ovary with the preparation of the wall of the uterus for the fertilised egg. It is usual to describe the cycle as starting on the first day of **menstruation**, the period of bleeding from the uterus. This is the stage when the thickened lining of the uterus wall breaks down. The adaptive advantage of this process has long been debated and there is no clear answer. It seems wasteful to expend much energy in building a thick new layer each month and then disposing of it. One plausible suggestion is that, if left in place, the layer would slowly deteriorate and be less satisfactory for the embryo to implant. It is likely that the survival rate of embryos is increased by the regular renewal, so natural selection has favoured this system.

Four different hormones are involved in the menstrual cycle, two made in the ovaries and two produced by the pituitary gland. Together these hormones coordinate the maturing and release of an ovum and the rebuilding of the uterus wall. Figure 26 enables you to compare the amounts of each hormone produced with the events in the ovary and the uterus.

During the first week of the cycle the pituitary gland is stimulated to secrete the hormone **FSH**, which is short for **follicle stimulating hormone**. FSH stimulates one of the potential egg cells in one ovary to develop. As well as the ovum itself, the cells surrounding it grow as you can see from Figure 26. This is called a **follicle**. The follicle cells secrete **oestrogen**. Oestrogen stimulates the rebuilding of the uterus wall. You can see from the diagram that this starts as soon as the old layer has been shed on about day 5 of the cycle.

At first the amount of oestrogen produced is quite low. This low concentration in the blood has a negative feedback effect on the secretion of FSH by the pituitary gland. You can see from the graph that after about day 7 of the cycle the concentration of FSH decreases again. Figure 27 shows this negative feedback effect.

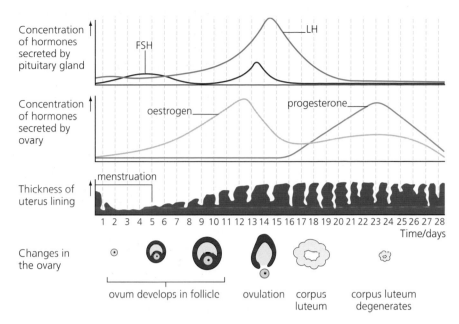

▲ **Figure 26** The menstrual cycle in humans.

As the follicle grows it produces larger quantities of oestrogen and the concentration in the blood rises. When it reaches a threshold, its effect on the pituitary gland reverses. It changes to having a positive feedback effect on the secretion of FSH so there is a surge in the amount of FSH produced. It also stimulates the pituitary gland to release **LH** (**luteinising hormone**) (Figure 28). The graph in Figure 26 shows that there is a steep increase in the amounts of both FSH and LH on days 12 and 13.

The peak of LH concentration stimulates the follicle to burst and release the ovum about 9 hours later, on about day 14 of the cycle. The follicle does not disappear completely, but reforms into a structure called the **corpus luteum**. LH stimulates the corpus luteum to produce **progesterone**. This

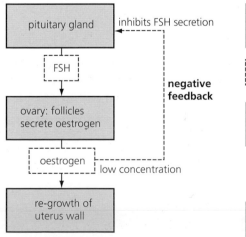

▲ **Figure 27** Negative feedback effect of low oestrogen level on FSH.

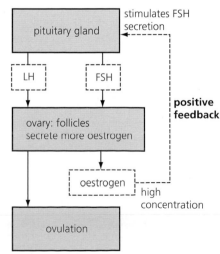

▲ **Figure 28** Positive feedback effect of high oestrogen level on FSH, and the effect on LH.

hormone completes the development of the uterus wall by increasing the blood supply and promoting glycogen storage. The wall is then fully prepared to receive an embryo. The rising concentrations of progesterone and oestrogen have a negative feedback effect on LH and FSH production. The fall in FSH concentration prevents any new follicles developing.

If fertilisation is to be successful it normally must occur within two days of ovulation. It takes the developing embryo another three days or so to reach the uterus and implant in the wall. If fertilisation does not take place and no embryo implants within about a week, the corpus luteum starts to degenerate. What causes this is not fully understood. It seems to be in part the result of the declining LH concentrations, but it is thought that prostaglandins are also involved (page 189). As progesterone and oestrogen concentrations decline, the breakdown of the uterus wall begins. FSH secretion increases and the menstrual cycle starts again.

Of course if an ovum is fertilised and an embryo is implanted, it would be disastrous for the wall to be broken down and ejected. This is prevented by another hormone (hCG) produced at the site of implantation. This stops the degeneration of the corpus luteum and sets up the complex sequence of events involved in pregnancy. The presence of hCG in the blood can be detected at quite an early stage. It is on this basis that pregnancy testing kits work. A complex set of hormonal interactions regulates the growth and development of the embryo and fetus until the birth of a baby some 9 months later.

Chapter 10

How genes work and their control

Figure 1 shows three stages in the life cycle of a frog. The tadpoles live in water. They have gills and a tail, which adapt them for life in this environment. The young tadpole initially feeds on yolk, stored in the egg from which it hatched, before becoming a plant eater. The older tadpole is a carnivore. Notice the external gills of the young tadpole which you can no longer see in the older tadpole because they are inside its body. The adult frog lives mainly on land. It has neither a tail nor gills; it uses its skin and lungs as a gas-exchange surface.

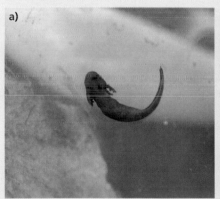

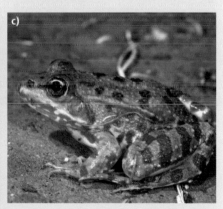

▲ **Figure 1** Three stages in the life cycle of a frog: **a)** young tadpole with external gills; **b)** older tadpole with internal gills; **c)** adult frog.

You might be familiar with other animals that have different body forms during their life cycle. The cabbage white butterfly shown in Figure 2 is a major economic pest in the UK. Its larval stage, the caterpillar, is a voracious leaf-eater and can destroy cabbage crops. The adult butterfly does not eat plant leaves; instead it feeds infrequently on the sugary nectar found in flowers.

Even if you know little about plants, you are likely to have seen ivy growing on walls or in woodland. Figure 3 shows part

▲ **Figure 2** You might have seen cabbage white butterflies in your garden. The larva **a)** feeds on brassica (cabbage) leaves and is completely different from the adult **b)**.

▲ **Figure 3** Ivy plants are common on garden walls. How many plant organs can you identify in this photograph?

of an ivy plant. In this photograph, you can see three different plant organs – leaves that photosynthesise, a stem that supports the leaves and transports substances between roots and leaves, and roots growing from the stem that hold the plant on to the wall and absorb water and inorganic nutrients. Also, not visible here, the leaves take different forms – the immature leaves lower down the stem have a different shape from the mature leaves towards the tip of the stem.

In all these examples, we can see different body forms within a single species. Body form is one aspect of an organism's phenotype. Although frogs and butterflies have different phenotypes during their life cycle, their genotype is the same at each stage. Similarly, the genotypes of the root, stem and differently shaped leaves of a single ivy plant are identical.

Have you ever taken cuttings from a plant and used them to grow new plants? If so, you have created a plant **clone** – a group of genetically identical plants. Humans have been doing this for hundreds, if not thousands, of years. Research botanists perform a more sophisticated version of the same process when they isolate individual cells from the root tip of a plant and grow them in tissue culture to produce large numbers of genetically identical plants.

Making clones from animals is not as easy. As you will see in this chapter, one reason for this is that animal cells lose the ability to express many of their genes as they differentiate and mature. However, some animals have been cloned by removing the nucleus from one of their cells and inserting it into an egg cell whose nucleus has been removed (Figure 4). This technology is called **somatic cell nuclear transfer** (**SCNT**); we usually refer to it as animal cloning.

You are probably familiar with Dolly, shown in Figure 5. She was produced by a team of scientists in Scotland, using somatic cell nuclear transfer, and is thought to be the first mammal to be successfully cloned. Dolly was put down when she was only six years old. Although she had given birth to a number of healthy lambs, at the time of her death Dolly

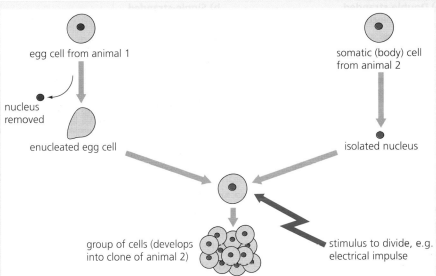

▲ **Figure 4** During somatic cell nuclear transfer (SCNT) the nucleus of a fertilised egg cell is removed and then replaced with the nucleus from a somatic cell from an animal of the same species.

showed symptoms seen more commonly in much older sheep. These included arthritis, lung disease and obesity. Some of these symptoms are thought to have resulted from problems associated with cell division and the control of gene expression in her body cells.

▲ **Figure 5** Dolly the sheep was the first mammal to be produced by somatic cell nuclear transfer. Do you know why she was called Dolly?

In this chapter, you will see how the genetic information carried in genes is used and you will learn how genes are controlled at a cellular level – how they are switched on or off in different cells within an organism. This knowledge will help to explain how different body forms within one life cycle and how different organs within one body are possible.

DNA and RNA

An organism's genes are made of DNA. You learnt about the structure of DNA in your AS course. However, before you can understand how genes work, you need to learn about another type of nucleic acid, called **ribonucleic acid (RNA)**.

Like DNA, a molecule of RNA is a **polynucleotide chain**. Figure 6 shows a DNA nucleotide, which you are already familiar with, and an RNA nucleotide. Can you spot the difference? Look at carbon atom 2 of the 5-carbon sugar. In the DNA nucleotide, it lacks an oxygen atom that is present in the RNA nucleotide. There is another difference that is not shown in Figure 6. An RNA nucleotide never has thymine as its base; instead it has uracil.

Figure 7 overleaf shows the structure of DNA and of two different types of RNA – messenger RNA (**mRNA**) and transfer RNA (**tRNA**). How many differences in the structures and compositions of DNA, mRNA and tRNA can you spot in Figures 6 and 7? Table 1 summarises these differences for you.

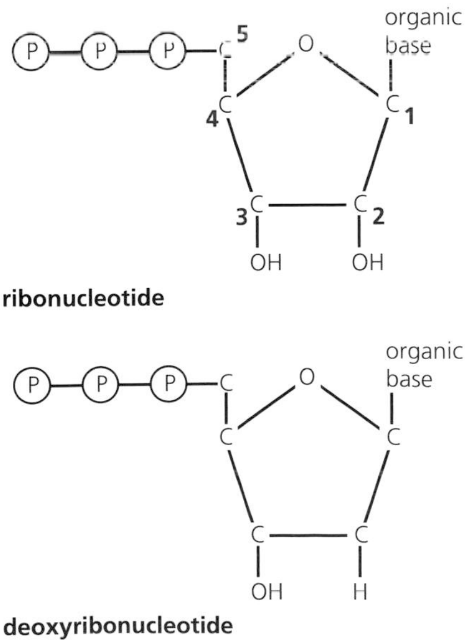

ribonucleotide

deoxyribonucleotide

▲ **Figure 6** The molecular structures of a RNA nucleotide and a DNA nucleotide. RNA is made up of the sugar ribose; DNA is made up of the sugar deoxyribose (which has one less oxygen than ribose).

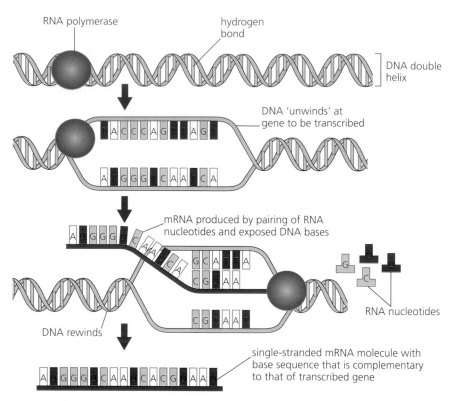

▲ **Figure 9** During transcription, the base sequence of DNA is copied into the complementary base sequence of messenger RNA. A gene is thus copied for use by ribosomes.

Q3

During transcription, the base sequence of only one of the strands of DNA is used to make a molecule of mRNA; the other DNA strand is not used. Give **one** advantage of a molecule of DNA having both strands.

In this way, a chain of RNA nucleotides is made which has a complementary base sequence to the DNA making up one gene. An enzyme called **RNA polymerase** joins the ribose-phosphate backbone of these RNA nucleotides to form a molecule of RNA. This RNA molecule is called **messenger RNA** (**mRNA**, for short).

Post-transcriptional processing of mRNA in eukaryotic cells

The process of transcription described above is fine in prokaryotic cells; all of their DNA codes for polypeptides. However, much of the DNA in eukaryotic cells does not code for polypeptides. The non-coding sections of DNA might be:

- between genes: these sections include DNA sequences that are repeated over and over again. They are often referred to as tandem repeat sequences
- within genes: these non-coding sections of DNA are called **introns** and they separate the coding sequences called **exons**. It might help you to remember the difference between exons and introns if you think of **ex**ons as **ex**pressed sections of DNA and **int**rons as **int**ervening sections of DNA.

During transcription, eukaryotic cells cannot transcribe only the coding sections. Instead the whole gene, including introns and exons, is transcribed into a molecule called **pre-mRNA**. Before it leaves the nucleus, this pre-mRNA is edited. Figure 10 shows how this is done by removing the non-coding sections of the pre-mRNA. The coding sections are then spliced together to produce mRNA that carries only the coding regions of the gene.

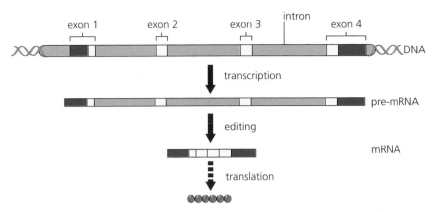

exon 1   exon 2   exon 3   intron   exon 4

DNA

transcription

pre-mRNA

editing

mRNA

translation

▲ **Figure 10** During transcription in a eukaryotic cell, the entire base sequence of a gene is copied into a base sequence of mRNA. This produces a molecule of pre-mRNA that contains a copy of the non-coding regions of the DNA (introns) as well as the coding regions (exons). The non-coding regions are edited out before the mRNA leaves the nucleus.

Translation

A new molecule of mRNA moves from the DNA to ribosomes in the cytoplasm. In a eukaryotic cell, this involves leaving the nucleus through one of the **nuclear pores** in the envelope surrounding the nucleus. One or more ribosomes attaches to a molecule of mRNA. Each ribosome moves along the molecule of mRNA, 'reads' the base sequence of the mRNA and uses it to assemble a sequence of amino acids. Peptide bonds form between the amino acids, forming a polypeptide. Once this polypeptide is complete, the ribosome releases the polypeptide and detaches from the molecule of mRNA.

A ribosome 'reads' the mRNA base sequence three bases at a time, i.e. 'reads' base triplets. Each base triplet is the genetic code for a specific amino acid. For this reason, each triplet of mRNA bases is called a **codon**.

Although amino acids can be free in the cytoplasm of a cell, they cannot be used by ribosomes unless they are attached to a molecule of transfer RNA (**tRNA**, for short). At the end of one of its arms, each tRNA molecule has three free bases that can pair with a complementary codon on mRNA (Figure 7, page 216). At the end of another of its arms, each tRNA molecule has a site that attaches to an amino acid. The sequence of bases on the tRNA molecule that pairs with a codon is called an **anticodon**. A molecule of tRNA with a particular anticodon always attaches to only one type of amino acid.

Figure 11 on the next page shows a single molecule of mRNA at an early stage of translation. It is attached to a ribosome and three of its bases, a codon, are shown within the ribosome. A molecule of tRNA is also shown. Notice that the tRNA has an anticodon that is complementary to the codon. The amino acid that is carried by the tRNA molecule is the one encoded by the base triplet on the mRNA.

Figure 12 shows a molecule of mRNA at a later stage of translation. In this drawing, you can see that the ribosome has moved along the mRNA molecule and has joined more amino acids together, so that a polypeptide chain is being formed by the ribosome.

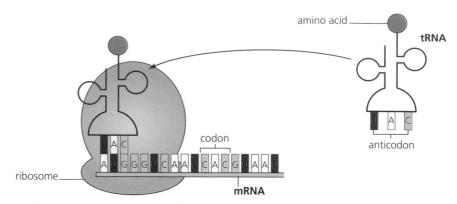

▲ **Figure 11** A ribosome has attached to one end of a molecule of mRNA. The ribosome is ready to 'read' the first three bases – the mRNA codon. You can also see a molecule of tRNA that has a base sequence that is complementary to the codon; this is the anticodon of the tRNA molecule. The tRNA is attached to an amino acid.

Q_4

a) What type of bond joins amino acids together in Figure 12?

b) What type of reaction is involved in joining amino acids together?

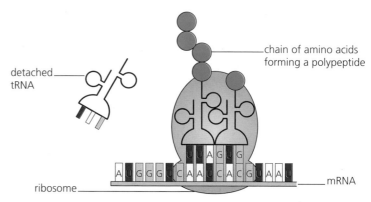

▲ **Figure 12** The ribosome has moved further along the mRNA molecule and has formed a chain of amino acids that corresponds to the codons along the mRNA molecule.

Once a ribosome has 'read' the entire base sequence of a molecule of mRNA, it releases the polypeptide it has formed. The ribosome then detaches from the mRNA. It might re-attach at the start of the mRNA molecule and produce another molecule of the polypeptide. Several ribosomes can attach to, and move along, a single mRNA molecule at the same time.

Alternative splicing

Cell biologists are able to analyse the base sequence of genes and the base sequence of mRNA molecules transcribed from them. They have found that only a small proportion of the total mRNA transcribed in human cells becomes mature mRNA that is used in translation.

1 Explain this observation.

One group of Polish scientists found the presence of non-coding base sequences even in mature mRNA. Some of these non-coding base sequences were trinucleotide repeats. These scientists found that trinucleotide repeats in one particular gene was associated with Huntington's disease. This disease is associated with deterioration of the nervous system.

2 Explain the term 'trinucleotide repeats'.

3 Suggest one reason why trinucleotide repeats might result in a disease.

4 Does the information confirm that Huntington's disease is caused by trinucleotide repeats? Explain your answer.

A different team of scientists investigated the way in which exons are spliced together during post-transcriptional processing. They found that exons can be spliced in many alternative ways to form different mature mRNA molecules. For example, early in their development human B cells splice into one of their mRNA molecules an exon that enables a particular protein to be retained in the cell's surface membrane. Later, they stop splicing this exon and, instead, splice into their mRNA an exon that enables the protein to be released from the cell.

5 Use your knowledge of immunology to suggest an advantage to a B cell of the above method of alternative splicing of exons into mature mRNA.

Further research has shown a dramatic example of alternative splicing in the fruit fly. One of its genes, called *Dscam*, is involved in a number of processes, including control of the growth of nerve cells and the recognition and phagocytosis of pathogens. This gene contains 116 exons, of which only 17 are ever retained in the mature mRNA. Some of these exons are always included in the mature mRNA but others are not. In theory, this alternative splicing could result in the production of 38 016 different proteins from the *Dscam* gene. In fact, scientists have found only about 18 000 in the blood of fruit flies.

6 Suggest why less than half the proteins that could result from alternative splicing of the *Dscam* gene have been found in the blood of fruit flies.

7 One of the discoverers of the structure of DNA called the 'one gene – one polypeptide' hypothesis the central dogma in biology. To what extent do you consider this is true?

The genetic code is universal, degenerate and non-overlapping

The genetic code is a sequence of bases in a DNA molecule. During the production of a polypeptide, the genetic code of a single gene is transcribed from DNA to a molecule of mRNA. This sequence of bases in mRNA is translated into a sequence of amino acids in a polypeptide. For this reason, the genetic code is given as mRNA codons. Table 3 on the next page summarises the genetic code. The code is the same for all organisms – it is **universal**. We can use the table to show two further important features of the genetic code.

Understanding how the genetic code is degenerate and non-overlapping

Look at the codons UAA, UAG and UGA in Table 3. These are 'reading instructions' for ribosomes. When a ribosome gets to these codons it detaches from mRNA and releases its polypeptide chain.

■ Is each amino acid encoded by only one codon?

Look at the codons UCA, UCG, UCC and UCU. They all code for the same amino acid – Ser (serine). This means that it does not matter about the third base; UCX always codes for the amino acid serine. We describe this by saying that the genetic code is **degenerate** – several mRNA

AAA	Lys	CAA	Gln	GAA	Glu	UAA	STOP
AAG	Lys	CAG	Gln	GAG	Glu	UAG	STOP
AAC	Asn	CAC	His	GAC	Asp	UAC	Tyr
AAU	Asn	CAU	His	GAU	Asp	UAU	Tyr

ACA	Thr	CCA	Pro	GCA	Ala	UCA	Ser
ACG	Thr	CCG	Pro	GCG	Ala	UCG	Ser
ACC	Thr	CCC	Pro	GCC	Ala	UCC	Ser
ACU	Thr	CCU	Pro	GCU	Ala	UCU	Ser

AGA	Arg	CGA	Arg	GGA	Gly	UGA	STOP
AGG	Arg	CGG	Arg	GGG	Gly	UGG	Trp
AGC	Ser	CGC	Arg	GGC	Gly	UGC	Cys
AGU	Ser	CGU	Arg	GGU	Gly	UGU	Cys

AUA	Ile	CUA	Leu	GUA	Val	UUA	Leu
AUG	Met	CUG	Leu	GUG	Val	UUG	Leu
AUC	Ile	CUC	Leu	GUC	Val	UUC	Phe
AUU	Ile	CUU	Leu	GUU	Val	UUU	Phe

Table 3 The genetic code. Three bases (a codon) on a molecule of mRNA encode a specific amino acid. The four bases, adenine (A), cytosine (C), guanine (G) and uracil (U) can form the 64 different codons that are shown in the table. Abbreviations for the names of the amino acids that they encode are shown alongside the codons. You do not need to remember the content of this table!

triplets encode the same amino acid. How many other amino acids can you find in Table 3 that are encoded by more than one codon? Can you find any that are encoded by only one codon?

Having learnt that some codons stop translation by a ribosome, you might wonder how a ribosome 'knows' where to start translating the mRNA code. This is slightly more complicated than the 'STOP' codons in Table 3. In most organisms, the ribosome must 'recognise' an AUG codon that is followed by a G and has an A preceding it by three nucleotides, for example the sequence AXXAUGG. The points you need to understand are that:

• some mRNA base sequences are start and stop messages for translation
• ribosomes only translate the mRNA molecule by taking a whole sequence of three bases at a time.

■ **Why is it important to understand that ribosomes only translate the mRNA molecule by taking a whole sequence of three bases at a time?**

You might realise that the sequence of mRNA bases can be 'read' in a variety of ways. It is important that the ribosome 'reads' the sequence in only one way, so that a polypeptide with the encoded sequence of amino acids is the only one that is produced.

To explain why, let's take the following mRNA base sequence:

UCCCAUGACUCAUUCCCAGGG

■ Use the information in Table 3 to work out the sequence of amino acids that will be produced by the mRNA base sequence above.

If translation starts at the first codon (UCC), the order of encoded amino acids will be

serine – histidine – aspartic acid – serine – phenylalanine – proline – glycine.

This results in the correct polypeptide being produced.

■ Suppose the ribosome missed the first U in the base sequence and started by translating the triplet CCC as the first codon. What amino acid sequence would now be produced?

If the ribosome missed the first mRNA base, it would now produce a polypeptide with the amino acid sequence
proline – methionine – threonine – histidine – serine – glutamine.

In other words, a completely different amino acid sequence would be produced, resulting in a polypeptide that would not have the appropriate function. Having a specific 'START' recognition sequence ensures that the mRNA code is always translated from the correct point in the mRNA molecule.

■ Now suppose that the ribosome were to 'read' the first mRNA base sequence above by jumping one base along each time it had 'read' a codon. What base sequence would it read?

The original base sequence would now be 'read' incorrectly as UCC, then CCC, CCA, CAU, AUG, and so on. This is an overlapping sequence and, again, would result in a polypeptide with an inappropriate sequence of amino acids.

Translating the first three bases as a single unit, followed by the fourth, fifth and sixth in a second unit, and so on, ensures that the mRNA code is **non-overlapping**. Each base is 'read' only once in the codon of which it is a part. As a result of this non-overlapping nature of translation, an appropriate amino acid sequence is always produced from a molecule of mRNA.

Gene mutation

Complementary base pairing is essential if transcription and translation are to work properly. Table 4 summarises the pairing relationship between a base in a DNA nucleotide and the bases in complementary nucleotides of mRNA and then of tRNA.

Base on DNA nucleotide	Complementary base in the codon of a nucleotide of mRNA	Complementary base in the anticodon of a nucleotide of tRNA
Adenine	Uracil	Adenine
Cytosine	Guanine	Cytosine
Guanine	Cytosine	Guanine
Thymine	Adenine	Uracil

Table 4 Complementary base pairing is important during transcription and translation. This table shows the mRNA bases that are complementary with DNA bases and the tRNA bases that are complementary with mRNA bases.

Q5

a) Look at the middle two rows of Table 5 below. The final codon in the sequence is incomplete (GCx). Why can we safely identify Ala (alanine) as the encoded amino acid?
b) Use these two rows to explain the term 'frame shift'.
c) Use the bottom two rows in Table 5 to explain why a base substitution does not cause a frame shift.

DNA base sequences are copied during DNA replication. At a rate of about one in a million, errors occur during this copying process. Two of the errors that can occur are:

- **base deletion.** A base is lost from the DNA sequence. As a result, the whole base sequence following the deleted base changes. This is a called a **frame shift.** and results in a new sequence of amino acids after the deletion.
- **base substitution.** The 'wrong' base is included in the base sequence. This mutation might result in a different amino acid being included in the polypeptide chain. It does not cause a frame shift.

Table 5 shows the effects that can arise from base deletion and base substitution.

Original base sequence on mRNA	AG**A**	UAC	GCA	CAC	AUG	CGC
Encoded sequence of amino acids	Arg	Tyr	Ala	His	Met	Arg
mRNA base sequence after base deletion	AGU	ACG	CAC	ACA	UGC	GCx
Encoded sequence of amino acids	Ser	Thr	His	Thr	Cys	Ala
mRNA base sequence after base substitution	AGU	UAC	GCA	CAC	AUG	CGC
Encoded sequence of amino acids	Ser	Tyr	Ala	His	Met	Arg

Table 5 The first two rows of this table show the amino acid sequence encoded by part of a molecule of mRNA. The mRNA sequence is shown as individual codons to help you to read the code. The table demonstrates the effect on the encoded amino acid sequence of a deletion and of a substitution of the adenine shown in red.

Q6

a) Mutations in some cells are not important. Suggest why a mutation in the gene for haemoglobin is unimportant in a white blood cell.
b) Explain why a gene mutation can have important effects if it occurs in (i) a gamete and (ii) a gene controlling cell division.

Mutagenic agents

Natural mechanisms occur within cells that identify and repair damage to DNA. These mechanisms become ineffective if the rate of mutation increases above the normal, low rate. Many environmental factors increase the rate of mutation. They are called **mutagenic agents** and include:

- toxic chemicals, for example bromine compounds, mustard gas (used in a large number of conflicts to kill soldiers and civilians), peroxides

Q7

Some people are concerned that the aerials used to transmit mobile phone messages emit radiation that increases the rate of mutation. For this reason, they oppose the erection of these aerials in their neighbourhood. Suggest what you would need to measure in order to investigate whether these aerials increased the rate of mutation.

- ionising radiation, for example gamma rays and X rays
- high-energy radiation, for example ultraviolet light.

Gene mutations and cancer

Mitosis occurs during the cell cycle of eukaryotic organisms. The rate of mitosis is controlled by two groups of genes:

- proto-oncogenes, which stimulate cell division
- tumour suppressor genes, which slow cell division. These genes also promote programmed cell death (**apoptosis**) in cells with DNA damage that the cell cannot repair.

Gene mutations can occur in both these two types of gene. A mutated proto-oncogene, called an **oncogene**, stimulates cells to divide too quickly. A mutated tumour suppressor gene is inactivated, allowing the rate of cell division to increase.

An investigation into tumour formation in transgenic mice

Clones of laboratory mice are often used in investigations into the causes and effects of tumours.

1 What is a clone?

Scientists genetically transformed mice from a single clone to contain different oncogenes, called *myc* and *ras*. One group of mice contained only the *myc* oncogene, a second contained only the *ras* oncogene and a third contained both the *myc* and the *ras* oncogenes. The scientists then recorded the age at which the mice in each group developed tumours.

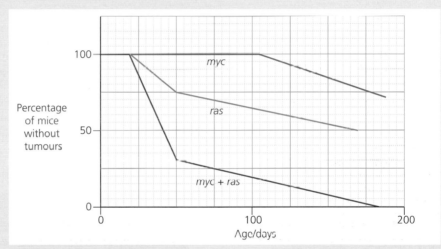

▲ **Figure 13** Three groups of mice from a single clone were genetically engineered to contain either *myc* oncogenes or *ras* oncogenes or both. The graph shows the percentage of mice in each clone that did *not* have tumours.

2 What is an oncogene?

3 Explain why the scientists used clones in this experiment.

Figure 13 shows the percentage of mice in each clone that were free of tumours during the course of the experiment.

4 For how many days were all the mice in this experiment free of tumours?

5 Use the graph to compare the effects of the *myc* and *ras* oncogenes when present alone.

6 Suggest an explanation for the curve showing mice with both the *myc* and *ras* oncogenes.

Figure 13 above shows the results of an investigation involving the formation of tumours in mice. A tumour is a group of one type of cell that is dividing rapidly and uncontrollably. The formation of a tumour might result from one, or only a few, genetic changes in a cell. Cancers are tumours in which some cells break away from the group and invade organs and tissues throughout the body. Figure 14 shows that the change from tumour cells to cancer cells requires many more genetic changes. Thus, cancer does not usually result from a single gene mutation.

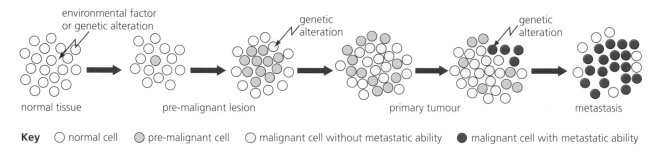

Key ○ normal cell ◔ pre-malignant cell ○ malignant cell without metastatic ability ● malignant cell with metastatic ability

▲ **Figure 14** Healthy cells become tumour cells as a result of mutation in one, or a few, genes controlling cell division. Many more mutations are required for a tumour cell to become a cancer cell.

Gene specialisation

We have seen how genes are transcribed and translated and some of the things that can happen when this process goes wrong.

We now look at the case of specialised cells. Specialised cells do not express all their genes. The human egg cell shown in Figure 15 is **totipotent**. This means that it is unspecialised and is capable of expressing all of its genes. The other specialised cells shown in Figure 15 are *not* totipotent; they have differentiated into mature tissue cells and have lost the ability to become totipotent again.

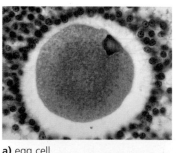

a) egg cell

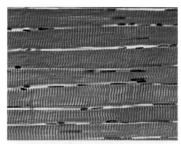

b) muscle cells

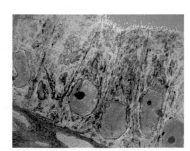

c) columnar epithelial cells

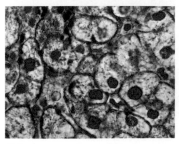

d) hepatocytes (liver cells)

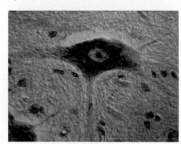

e) nerve cell

▲ **Figure 15** Each of these cells could be found in a single individual. Although they are genetically identical, the appearance, chemical composition and behaviour of these cells are different.

▲ **Figure 16** These apricot plantlets were grown *in vitro* from isolated somatic cells taken from parent plants. Plantlets from carrot cells have been grown in this way in space, on-board the un-staffed Soviet biosatellite Kosmos 782.

Q8

Suggest two processes controlled by housekeeping genes.

▲ **Figure 17** Bindweed is one of the top ten garden weed pests in the UK. If even a small part of its root is left in the soil, it will quickly grow to produce a new plant.

Q9

Embryonic mouse skin cells are used in culturing human embryonic stem cells to provide a surface to which the human cells can attach. Suggest one potential danger of this use of mouse cells.

Some genes are essential for the survival of cells. They are called **housekeeping genes** and are expressed in all the cells in an organism. Other genes are expressed only in a particular type of specialised cell or at a particular stage of development. For example, you express the gene encoding myosin in your muscle cells but not in your brain cells. Table 6 shows how you expressed different genes encoding one of the polypeptide chains of your haemoglobin molecules during your lifetime.

Stage of development	Gene encoding polypeptide chain of human haemoglobin that is expressed
Conception to about 6 weeks gestation	ξ
About 6 weeks gestation to birth	γ
Birth to adulthood	β

Table 6 The haemoglobin molecule of a human adult is made of two chains of α globin and two chains of β-globin. Different genes control the production of the β-globin chain at different stages of development.

Somatic cells are those cells in an organism that are not involved in reproduction. Many more somatic cells in mature plants retain the ability to divide and form new tissues than in mature animals. Given appropriate conditions, nutrients and growth substances, these totipotent plant cells can be encouraged to grow in tissue culture and develop into plant organs or into plantlets, such as the ones shown in Figure 16.

Totipotent plant cells will not only grow in tissue culture in laboratories – they will also do so naturally. Despite its attractive white flowers, the bindweed shown in Figure 17 is an unwelcome sight in most gardens. It wraps itself around other garden plants, competing with them for water, minerals and light. As it is in close contact with garden plants, it is difficult to kill using weedkillers because these will also kill the garden plants. The best way to kill bindweed is to pull it out of the soil. However, if any part of the root of the bindweed is left in the soil, it will quickly regenerate into a new plant. The root can do this because it contains cells that are totipotent – they are able to develop into any cells in the mature plant.

Stem cells

Unlike plants, only a small number of cells in animals retain the ability to divide and give rise to new tissues. They are called **stem cells** and can be found in embryonic tissue and in some adult tissues. Whatever their source, stem cells have three general properties.

- They can divide and renew themselves over long periods.
- They are unspecialised.
- They can develop into other specialised cell types.

Human embryonic stem cells are stem cells that have been taken from an embryo that has developed following *in vitro* fertilisation of an egg (in an *in vitro* fertilisation clinic) and been donated for research purposes with the informed consent of the donors. About 30 cells are removed from the inside of a human embryo that is 4 to 5 days old. These cells are plated into dishes that contain a coating of embryonic mouse skin cells and an appropriate culture medium, where they divide.

Very small numbers of **human adult stem cells** are found in specific tissues, such as bone marrow and the brain. They lie dormant for many years until stimulated to begin to divide, following an injury. In addition to being difficult to isolate, they are also more difficult to grow in tissue culture than embryonic stem cells. Unlike human embryonic stem cells, adult stem cells can only give rise to a limited number of body tissues – they are **multipotent**. In contrast, embryonic stem cells are **pluripotent** – if cultured in appropriate conditions, they can develop into most of the body's cell types.

The use of embryonic stem cells is controversial. In some countries, including the USA, it is currently illegal to use embryonic stem cells, even for research. In other countries, the use of embryonic stem cells is currently legal but is tightly regulated. In April 2008, members of the European Parliament (MEPs) voted to ban across the European Union any research involving embryonic stem cells. The following month, members of the UK parliament, which is a member of the European Union, voted to continue to allow such research.

Q10

Scientists hope that transplanting cultures of stem cells might be used to repair damaged tissues or to replace malfunctioning tissues.
a) How can transplanted stem cells repair or replace malfunctioning tissues?
b) Suggest one biological advantage and one biological disadvantage of using embryonic stem cells in such transplants.

Control can occur at several stages of gene expression

Gene expression involves the following flow of genetic information: DNA → mRNA → polypeptide. Gene expression can be controlled at any stage in this flow of information.

- Control of transcription: only some genes are transcribed.
- Control of post-transcriptional processing: the way in which different combinations of exons in the pre-mRNA are spliced into mature mRNA is controlled.
- Control of translation: mRNA might be destroyed or its translation by a ribosome blocked.
- Control of post-translational editing: polypeptides are often modified in the cytoplasm after they have been produced.

You need to learn about only two of these controls: control of transcription and control of translation.

Control of transcription by specific transcriptional factors

Every gene is surrounded by one or more DNA base sequences that control its expression. Figure 18 shows one of these, called a **promoter region**. A promoter region is located near the gene it controls, usually about 100 base

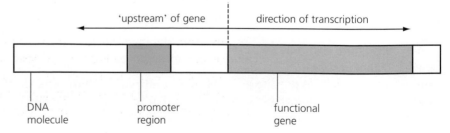

▲ **Figure 18** Every gene in eukaryotes is controlled by one or more promoter regions. The promoter region lies close to the gene it controls and is 'upstream' of it.

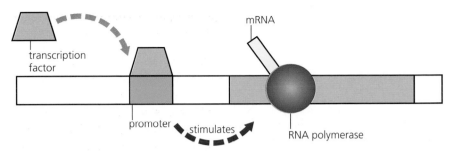

▲ **Figure 19** The role of a transcription factor and a promoter region in stimulating transcription. The position of a gene encoding a polypeptide is shown in part of a molecule of DNA. In the DNA sequence preceding this gene (referred to as 'upstream' of the gene) is a promoter region. Only if an appropriate transcription factor attaches at this promoter region can RNA polymerase begin to transcribe the gene.

pairs before the start of its gene (approximately 100 base pairs **upstream** of the gene). Figure 19 shows how a protein, called a **transcription factor**, binds to a gene's promoter region and, in doing so, enables RNA polymerase to attach to the start of the gene and begin its transcription.

Oestrogen enables RNA polymerase to begin transcription of its target gene

Oestrogen is a mammalian hormone. You have learnt about the role of oestrogen in the control of the mammalian oestrous cycle; it also stimulates sperm production in males. Because it has small, hydrophobic molecules, oestrogen can diffuse through the plasma membranes of cells. Once in the cytoplasm, oestrogen diffuses into the cell's nucleus, where it binds to a type of oestrogen receptor, called ER alpha (**ERα**).

The ERα oestrogen receptors are transcription factors that can bind to the promoter region of up to 100 different genes. Figure 20 on the next page shows one way in which these ERα oestrogen receptors work. In the cell, oestrogen receptors are normally held within a protein complex that inhibits their action. When oestrogen binds to an ERα oestrogen receptor it causes the oestrogen receptor to change shape and leave its protein complex. The oestrogen receptor can now attach to the promoter region of one of its target genes, stimulating RNA polymerase to transcribe the target gene.

Oestrogen receptors, oestrogen-dependent breast tumours and hospital budgets

You learnt about the causes of tumours, including cancers, on page 225. Many people in the UK develop breast tumours; the vast majority are women. About 35% of breast tumours are associated with over-stimulation of the gene encoding oestrogen. They are called **oestrogen-dependent breast tumours**.

For over 20 years, the main treatment of breast tumours has been a drug called tamoxifen. This drug is effective because it has a chemical shape similar to oestrogen (Figure 21). This enables tamoxifen to bind permanently to oestrogen receptors in tumour cells. As a result, these tumour cells can no longer bind with oestrogen, which they need to grow.

More recently, new drugs have been developed that inhibit an enzyme, aromatase, which is required for the synthesis of oestrogen. Like tamoxifen,

Q11

Explain why having hydrophobic molecules enables oestrogen to diffuse through the plasma membrane of a cell.

Q12

Oestrogen has widespread effects in humans, affecting the reproductive system, the cardiovascular system, the immune system and bone tissue. Use information in the text to suggest how this is possible.

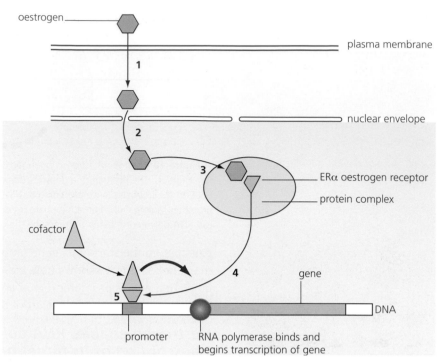

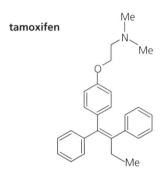

tamoxifen

oestrogen

▲ **Figure 21** Tamoxifen is a drug that is used to treat oestrogen-dependent breast tumours. Tamoxifen has a similar chemical shape to oestrogen.

▲ **Figure 20** A simplified summary of how oestrogen stimulates the transcription of a target gene. **(1)** Oestrogen diffuses through the plasma membrane of a target cell and then diffuses into its nucleus **(2)** Here it attaches to an ERα oestrogen receptor that is contained within a protein complex **(3)** This causes the oestrogen receptor to change its shape and leave the protein complex that inhibits its action **(4)** The oestrogen receptor can now attach to the promoter region of a target gene **(5)** where it attracts other cofactors to bind with it. The oestrogen receptor, with combined cofactors, enables RNA polymerase to transcribe its target gene.

these drugs stop tumour cells responding to oestrogen, so stopping their growth. Large-scale clinical trials suggest that aromatase inhibitors, such as anastrozole and letrozole, can be even more effective in treating breast tumours than tamoxifen.

However, not all breast tumours are stimulated by oestrogen. Tamoxifen and aromatase inhibitors are ineffective against breast tumours that are not oestrogen-dependent. A relatively new drug, called herceptin, is effective against some of these breast tumours because it controls them in a different way. Herceptin is a monoclonal antibody. You learnt about monoclonal antibodies in your AS course. Herceptin works by binding to a growth factor protein that is embedded in the plasma membrane of some types of breast-tumour cells. As a result, it inhibits the growth of the tumour cells. The type of breast-tumour cell against which herceptin is effective is very invasive. Treatment of patients with this type of breast tumour, used in conjunction with chemotherapy, has been shown to lead to long periods of disease-free remission.

However, there is one major drawback with the use of herceptin. It is very expensive, costing the National Health Service about £22 000 per patient per year at 2008 prices. National Health Trusts have to consider the cost of treating patients, since they have a limited budget and have to balance the cost-effectiveness of what they can do. Some Trusts have not been able to afford herceptin treatment for patients with invasive breast cancers. This has

Q13

If you were a board member of a National Health Trust involved in a decision about whether herceptin should be used to treat a patient with breast cancer, what type of evidence would you take into account?

▲ **Figure 22** Petunia flowers show a wide range of colours.

led to a so-called postcode lottery – depending on where a sufferer lives she might, or might not, be able to receive herceptin treatment. As is often the case, the use of biological advances is dependent on decisions made by members of society.

Control of translation by small interfering RNA (siRNA)

Petunias are popular plants that are grown in hanging baskets in the UK (Figure 22). The purple colour of the flowers in Figure 22 results from a series of reactions in which a white pigment is converted to a purple pigment. One of the enzymes in this series of reactions is called chalcone synthase.

white pigment → → → → → → purple pigment

In 1990, a group of scientists reported their attempts to produce petunia flowers with a very deep purple colour. They used genetic engineering techniques to insert many copies of the gene encoding chalcone synthase into the cells of petunia plants with pale purple flowers.

■ **Use your knowledge of how enzymes work to suggest why the scientists expected the plants with the extra genes for chalcone synthase to produce deep purple flowers.**

Enzymes speed up a reaction by combining with molecules of substrate to form enzyme-substrate complexes, which break down to release molecules of the product. We can increase the rate at which the product is formed by adding more enzyme molecules. As a result, more enzyme-substrate complexes will be produced and so the product will be formed faster. The scientists predicted that if they added more genes for chalcone synthase to cells of petunia plants, more mRNA would be transcribed from them and so more molecules of enzyme would be made in the cells of the petunia.

Instead of deep purple flowers, the transformed petunia plants produced flowers that were mottled white. The scientists could not explain this result.

In 2004, in a totally unrelated investigation, a different team of scientists reported their discovery of enzymes called RNA-dependent RNA polymerases (**RDRs**). These RDRs catalyse the production of double-stranded RNA (**dsRNA**). They do this by using molecules of mRNA that are in the cytoplasm as a template to synthesise a complementary strand. The two RNA strands are held together by hydrogen bonds between complementary base pairs.

■ **What is unusual about the RNA produced by RDRs?**

You have learnt earlier in this chapter (Table 1) that RNA molecules are single-stranded. The RNA molecules produced by RDRs are double-stranded (dsRNA). Figure 23 shows what happens when RDRs produce a dsRNA molecule from a molecule of mRNA in the cytoplasm.

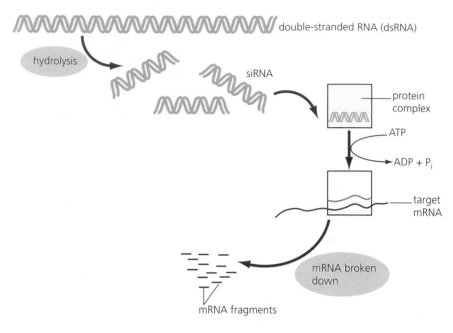

double-stranded RNA (dsRNA)

hydrolysis

siRNA

protein complex

ATP

ADP + P$_i$

target mRNA

mRNA broken down

mRNA fragments

▲ **Figure 23** Small interfering RNA (siRNA) are small double-stranded RNA molecules. They are used by protein complexes in a cell's cytoplasm to break down mRNA. By breaking down the target mRNA, a cell can control the expression of the gene from which the mRNA was transcribed.

■ **Look at Figure 23. What is the first thing that happens to the dsRNA molecule?**

The dsRNA molecule is cut into small fragments. These fragments are about 23 base pairs long and are called **small interfering RNA (siRNA)**. Figure 23 shows what happens to the siRNA. In a reaction requiring the hydrolysis of ATP, a protein complex in the cytoplasm takes up one of these siRNA fragments and separates its two RNA strands.

■ **Suggest why the hydrolysis of ATP is involved in the reaction between the protein complex and the siRNA it takes up.**

Like many reactions in metabolism, the reaction between the protein complex and the siRNA requires energy. This energy is released when ATP is hydrolysed to ADP and inorganic phosphate.

Look at the protein complex containing the single strand of RNA in Figure 23. The protein complex uses the RNA strand to bind to a molecule of mRNA in the cytoplasm. Once it has attached to the mRNA molecule, the protein complex breaks down the mRNA molecule. This stops the mRNA being translated by the cell's ribosomes.

■ **To which type of mRNA will the protein complex attach?**

The RDR in Figure 23 used a molecule of mRNA to make the dsRNA. One strand of the dsRNA must, therefore, have a complementary base sequence to the mRNA from which it was made. When the protein complex separates the two strands of the siRNA, it uses this complementary base sequence to attach to any of the original mRNA molecules in the cytoplasm. Thus, the protein complex specifically breaks down the mRNA from which the siRNA was made.

■ **Will the base sequence of the siRNA be complementary to the whole of the mRNA in Figure 23?**

The siRNA is a small fragment of the mRNA molecule, so it will only be complementary to part of the base sequence of the mRNA. However, this is enough to enable the protein complex to bind with, and destroy, the target mRNA.

■ **The destruction of mRNA by siRNA is stimulated when the concentration in the cytoplasm of one type of mRNA becomes high. Can you use this information to suggest how the discovery of RNA-dependent RNA polymerases and small interfering RNA provided an explanation for the failure of the first team of scientists to produce petunia flowers with a deep purple colour?**

The first team of scientists increased the concentration of mRNA encoding chalcone synthase by inserting into petunia cells many copies of the gene for the enzyme. A high concentration of mRNA encoding chalcone synthase stimulated RDRs to produce dsRNA from it. As a result many more siRNA molecules carrying part of the mRNA code for chalcone synthase were produced by the plants, so the mRNA was destroyed by the mechanism in Figure 23. Unwittingly, the scientists had stimulated destruction of mRNA encoding chalcone synthase instead of stimulating more of it to be transcribed to produce more enzyme molecules.

Chapter 11

Gene cloning and gene transfer

 Do you intend to study biology at university? If so, are you planning to use your degree to make a career? Many graduate biologists do – for example your biology teachers. Some biologists are employed by large pharmaceutical companies, where they are employed to develop new drugs.

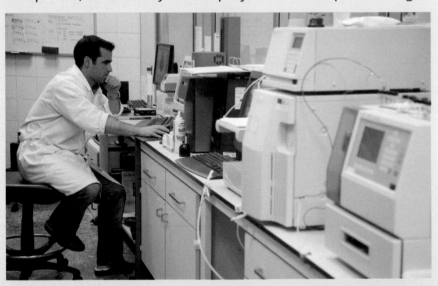

▲ **Figure 1** There are many careers open to biology graduates. Research and development is just one area.

In the 1980s, pharmaceutical companies were attracted by new developments in DNA technology, including the genetic manipulation of organisms. Some molecular biologists even set up their own companies to capitalise on their research. The potential profits from this biotechnology industry attracted people to invest their money in these companies. They believed that buying shares in these companies would give them a better return on their money than other types of investment. This sudden growth in ownership of shares in the biotechnology industry was called the 'biotechnology bubble'. Biotechnology companies became very valuable in a short space of time and some molecular biologists who had set up their own companies became paper millionaires.

Though the potential benefits of biotechnology were promising, there were many difficulties in converting research methods into profit-earning products. Many of these problems were technical, but some resulted from negative public opinion. As a result, the price of shares in these companies fell dramatically in 1999; the biotechnology bubble burst and investors lost a good deal of the money they had invested.

Molecular biologists work with fragments of DNA

Entire molecules of DNA are, in general, too large to be used in gene technology. Instead, molecular biologists use fragments of DNA, which contain the gene or genes they are interested in. They can produce these fragments using three different methods:

- Use a reverse transcriptase to convert mRNA to cDNA.
- Use a restriction endonuclease to cut DNA fragments out of a large DNA molecule.
- Use the polymerase chain reaction (PCR) to amplify part of a DNA molecule.

We will look at each of these methods in turn.

Using a reverse transcriptase to convert mRNA to cDNA

You learnt in Chapter 10, page 226, that most cells do not transcribe all their genes. Thus, the cytoplasm of a specialised cell only contains mRNA transcribed from some of the genes in its nucleus. It is often easier to extract mRNA from the cytoplasm of a cell than to find the gene from which it was transcribed. For example, mRNA encoding the mammalian hormone insulin will be found in high concentrations in the cytoplasm of β-cells in the islets of Langerhans of the pancreas (Chapter 9, page 205) but in no other cell.

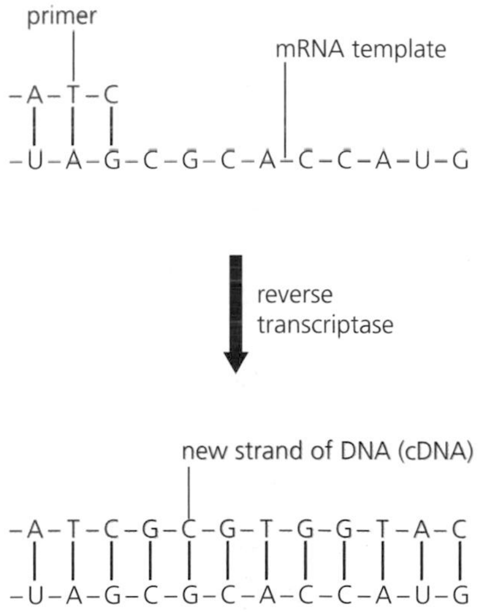

Molecular biologists can extract this mRNA and make a DNA copy from it (Figure 2). Purified mRNA is mixed with free DNA nucleotides and an enzyme called **reverse transcriptase**. This enzyme catalyses a process which is the reverse of transcription – DNA is made from mRNA. The DNA formed in this way is called **complementary DNA (cDNA)**.

Using a restriction endonuclease to cut fragments out of a large molecule of DNA

Nucleases are enzymes that break the bonds linking one nucleotide to the next in a DNA strand. An **exonuclease** removes nucleotides, one at a time, from the end of a DNA molecule. We are more interested in **endonucleases**. Figure 3 overleaf shows how an endonuclease breaks bonds within the DNA molecule, producing fragments of DNA.

Restriction endonucleases (or restriction enzymes, for short) are enzymes that break bonds in the sugar-phosphate backbones of *both* strands in a DNA molecule. As a result, they produce double-stranded fragments of DNA, like the ones shown in Figure 3. Each restriction enzyme cuts DNA only at a particular sequence of bases, called the **recognition sequence** of the enzyme.

Q1

What will be the base sequence of cDNA made from a section of mRNA with the base sequence: ACG CGA UCA UGA?

primer

mRNA template

$-A-T-C$
$\quad | \quad | \quad |$
$-U-A-G-C-G-C-A-C-C-A-U-G$

↓ reverse transcriptase

new strand of DNA (cDNA)

$-A-T-C-G-C-G-T-G-G-T-A-C$
$\quad | \quad | \quad | \quad | \quad | \quad | \quad | \quad | \quad | \quad | \quad | \quad |$
$-U-A-G-C-G-C-A-C-C-A-U-G$

▲ **Figure 2** Complementary DNA (cDNA) is made using reverse transcriptase to copy the base sequence of mRNA.

Q2

Reverse transcriptase can only begin to copy mRNA in the presence of a primer. Use Figure 2 to suggest what a primer is.

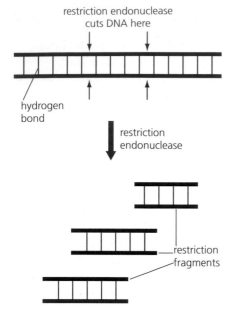

▲ **Figure 3** A restriction endonuclease cuts DNA into double-stranded fragments, called restriction fragments.

Q3

The recognition sequence of a certain restriction endonuclease occurs six times along a molecule of DNA. How many fragments of DNA will be produced by the action of the restriction enzyme?

A restriction endonuclease can cut a DNA molecule in one of two ways. Look at Figure 4 a). This shows how one restriction enzyme makes a simple cut right across the middle of its recognition sequence. You can see that this results in DNA fragments with **blunt ends**. Figure 4 b) shows how another restriction enzyme makes a staggered cut across its recognition sequence. This produces fragments with short single-stranded overhangs. Since these overhangs can form base pairs with other complementary sequences, they are called **sticky ends**.

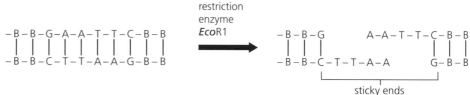

a) **Production of blunt ends by the restriction enzyme Alu1**

b) **Production of sticky ends by the restriction enzyme EcoR1**

▲ **Figure 4** A restriction endonuclease produces double-stranded fragments of DNA by breaking DNA molecules at a specific recognition sequence. The bases involved in the recognition sequence are shown in red. A is adenine, C is cytosine, G is guanine and T is thymine. The symbol 'B' represents other organic bases that are not involved in the recognition sequence.

The properties of recognition sequences

As well as showing the difference between blunt and sticky ends, Figure 4 shows some other features of restriction endonucleases and of recognition sequences.

The odd names of restriction enzymes are derived from the bacteria from which they were isolated. *Alu*1 was isolated from *Arthrobacter luteus*; *Eco*R1 was isolated from *Escherichia coli*.

■ **What does Figure 4 tell us about the recognition sequence of the restriction enzymes from these two bacteria?**

The recognition sequence of *Alu*1 is different from the recognition sequence of *Eco*R1. In fact, every restriction enzyme has a unique, specific recognition sequence where it cuts a DNA molecule.

■ **Can you use your knowledge of how enzymes work to explain why each restriction enzyme has a *specific* recognition sequence?**

You should recall that enzymes are specific. An enzyme is specific because only one type of substrate can fit into its active site. The active sites of restriction endonucleases are specific to particular sequences of bases, i.e. to specific recognition sequences.

■ **How long are the recognition sequences in Figure 4?**

In Figure 4 a), the recognition sequence is only four base pairs (4 bp) long. In Figure 4 b), the sequence is only 6 bp long. Although some restriction enzymes have recognition sequences that are longer than this, most have recognition sequences between 4 bp and 8 bp long.

■ **Look carefully at the sequence of bases in the upper and lower strands of each DNA fragment shown in Figure 4 a) and b). What is the common feature about their patterns of bases?**

In Figure 4 a), the upper strand of the recognition sequence reads AGCT. If you read the lower strand from right to left, it also reads AGCT. Look again at Figure 4 b). The upper strand of the recognition sequence reads GAATTC, which is the same as the lower strand when read from right to left. The order of bases in one strand of a recognition sequence is always the same as the complementary strand when read backwards. Something that reads the same forwards and backwards is called a palindrome. A simple palindrome in English is 'never odd or even'. Figure 4 shows us that, in addition to being short, recognition sequences are palindromic. We can now define the recognition sequence of a restriction enzyme as a short, specific, palindromic base sequence.

■ **How could molecular biologists ensure that they only cut a single gene out of DNA?**

Provided they know the base sequences at each end of a gene, biologists can use restriction endonucleases with base sequences that are complementary to them. As a result, the two restriction enzymes will cut the DNA, one at each end of the desired gene. Of course, this will only be helpful if the recognition sequences do not also occur within the gene.

■ **Bacteria can be infected by viruses that inject their nucleic acid into the host cell. Suggest one natural function of restriction endonucleases in bacteria.**

A bacterium uses its restriction endonucleases to break down the nucleic acid injected by a virus before it can be transcribed. In this way, bacteria protect themselves against viral infection. The bacterium's own DNA is not broken down because, unlike the DNA of the virus, bacterial DNA contains methyl groups ($-CH_3$) that protect it from the action of its own restriction endonucleases.

Analysing restriction fragments

Since they are produced by restriction endonucleases, the DNA fragments produced by these enzymes are called **restriction fragments**. DNA is digested by restriction endonucleases into a number of restriction fragments of different lengths. We can find out the number and size of the restriction fragments using **gel electrophoresis** followed by visualisation of the DNA. You have come across these techniques before in your AS course.

Separating restriction fragments using electrophoresis

Electrophoresis uses a slab of gel, made of agarose or polyacrylamide. An electrode is placed at each end of the gel and an electric current is passed through it. A sample containing restriction fragments is placed in a well that is cut in the gel near the negative electrode. Since DNA has a negative charge, the restriction fragments will migrate through the pores in the gel to the positive electrode. The smaller fragments will move faster through the pores in the gel than the larger molecules and so will move the furthest from the well. Figure 5 shows a cross section through a gel with restriction fragments migrating through it. The fragments produce bands of restriction fragments of different sizes.

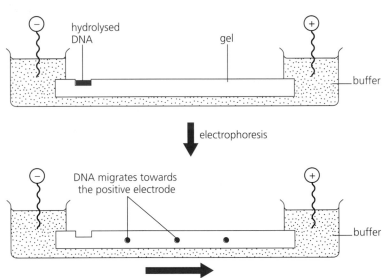

▲ **Figure 5** Restriction fragments migrate through a gel towards the positive electrode of an electric field. The smaller fragments move through pores in the gel faster than the larger fragments. Thus, bands of restriction fragments are formed in the gel.

Visualisation of DNA bands

DNA is colourless, so we must treat the DNA in such a way that we can see the bands in the gel after electrophoresis. This can be done using a suitable stain. The result is a series of coloured bands in the gel. An alternative is to treat the DNA with a **radioactive marker** before it is digested by restriction endonucleases. The fragments can then be seen by placing the gel in contact with a photographic film that is sensitive to X-rays. The developed film, called an **autoradiograph**, shows the DNA bands. Figure 6 opposite shows a developed film showing labelled restriction fragments.

Southern transfer

The gels used in electrophoresis are fragile. We can make a less fragile copy by transferring the pattern of bands from the gel to a more resilient membrane. Figure 7 shows a simple way of doing this. The membrane is placed on the gel and covered with lots of paper towels. A weight is placed on top of the paper towels and the apparatus is left for some time. During this time, the DNA bands are transferred from the gel to the membrane. We now have a more robust copy of the DNA bands that were present in the gel. As this technique was devised by a biologist called Professor E M Southern,

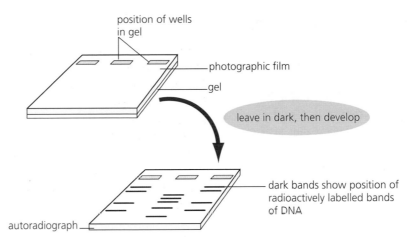

▲ **Figure 6** Radioactively labelled DNA has been separated using gel electrophoresis. After separation by electrophoresis, the gel has been placed in contact with a photographic film. After development, the different restriction fragments appear as dark bands on the film, called an autoradiograph.

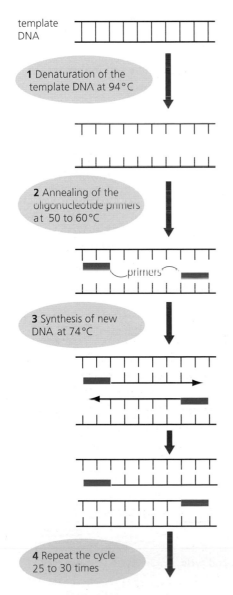

▲ **Figure 8** The basic steps in the polymerase chain reaction (PCR).

the technique is called a **Southern transfer** (also known as a **Southern blot**). Similar transfers can be made of the separated fragments of digested RNA and of digested proteins. (As a biological joke, these are called northern blots and western blots, respectively. No one has yet come up with an eastern blot!)

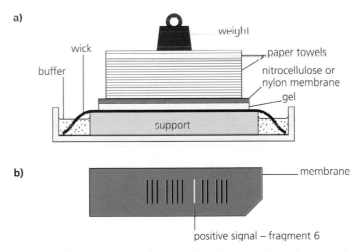

▲ **Figure 7** The pattern of DNA bands is transferred from an electrophoresis gel to a more resilient membrane, to form a Southern blot; **a)** shows a simple way of transferring the DNA bands to the membrane; **b)** shows the membrane with its copy of the DNA bands.

Using the PCR to amplify part of a DNA molecule

According to scientific folklore, the idea of the polymerase chain reaction (PCR) occurred as a brainwave to its inventor one evening as he drove along the coast of California. What is true is that the PCR has since become one of the most useful tools in molecular biology (Figure 8). It is highly likely that you have seen the PCR 'performed' in TV crime programmes.

The **polymerase chain reaction** is a rapid and simple procedure that is used to **amplify DNA**, i.e. to make multiple copies of DNA. It involves

Q4

The enzyme most commonly used in the PCR is *Taq* **DNA polymerase**. This enzyme is extracted from a bacterium, *Thermus aquaticus*, that lives in hot-water springs. Many of this bacterium's enzymes, including *Taq* DNA polymerase, are thermostable.

a) What is the advantage to the bacterium of having thermostable enzymes?

b) Explain why it is important that the DNA polymerase used in the PCR is thermostable.

Q5

You have learnt already about the importance of molecular collisions in explaining the action of competitive inhibitors. Use this understanding to suggest why it is important to the success of Step 2 that the PCR mixture contains a high concentration of primer.

mixing the DNA to be copied (**template DNA**) with a set of reagents in a test tube and placing the tube in a thermal cycler. The reagents include an enzyme, short lengths of single-stranded DNA (called **oligonucleotides**) and free DNA nucleotides, each with an adenine, cytosine, guanine or thymine base. The thermal cycler is an automated machine that varies the temperature at which the test tube is incubated in a pre-programmed sequence. Figure 8 shows the three basic steps in one PCR cycle, which are described in more detail below.

- **Step 1.** Denaturing of the template DNA. The thermal cycler heats the mixture to 94 °C. This causes the hydrogen bonds that hold together the two strands of the template DNA to break down. As a result, the two individual DNA strands separate and can now act as templates for building complementary strands.

- **Step 2.** Annealing the primers. The cycler cools the mixture to between 50 °C and 60 °C. This allows hydrogen bonds to re-form. In theory, the two strands of template DNA could re-join at this temperature. They do not because the oligonucleotides bind to the template DNA strands instead. This binding is called **annealing** and occurs only at a site on the template DNA that has a base sequence complementary to that of the oligonucleotides. These oligonucleotides are called **primers** because the enzyme involved in the next step must attach to one of them before it can start to work.

- **Step 3.** Synthesis of new DNA. The cycler raises the temperature to 74 °C. This is the optimum temperature of the **DNA polymerase** used in this step of the PCR. The enzyme attaches to one end of each primer and synthesises new strands that are complementary to the template DNA. At the end of this stage, the original molecule of template DNA has been copied and two molecules are in the mixture. The cycler now raises the temperature back to 94 °C and the cycle occurs all over again.

The PCR is usually repeated about 25 times. As a result, over 50 million copies of the template DNA are formed. The result of a small number of cycles is shown in Figure 9.

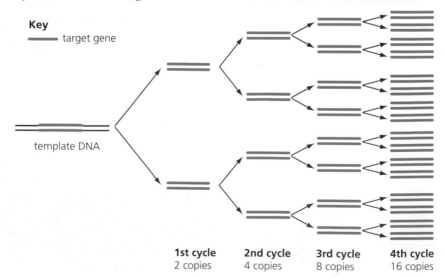

Figure 9 For each cycle, the PCR produces two molecules of DNA from a template molecule of DNA. In a short time, many copies of the template DNA can be made.

How do scientists use the PCR to amplify specific fragments of DNA, including genes?

So far, our account of the PCR suggests that the whole of a DNA molecule will be copied in each cycle. In eukaryotes, this would involve copying entire chromosomes, which would not be helpful.

■ Why would it be unhelpful to copy entire chromosomes?

Molecular biologists usually want to amplify only part of a DNA molecule, for example a gene that they wish to transfer from one organism to another or a gene that they wish to store in a gene library. A whole chromosome contains too much information to transfer or to store.

During the PCR, DNA polymerase starts replication where it binds to a primer. It will stop when it reaches another primer on the same strand of DNA. The correct design of primers is critical to the success of amplifying fragments of DNA.

■ It would be possible to design a primer that contained only two nucleotides. Why would this not be useful in the PCR?

The primer is a sequence of nucleotides that attaches to template DNA wherever it meets a base sequence that is complementary to its own. A sequence of only two bases, for example CT, is likely to occur a large number of times within a short length of template DNA. DNA polymerase starts copying when it attaches to a primer, then moves along the template DNA but stops when it reaches another primer. With primers attached only short distances apart, this would result in a large number of very small sections of amplified DNA.

It is easy to estimate how often base sequences of different lengths are likely to occur along template DNA. For example, a particular sequence of 4 bases is likely to occur once in every $4^4 = 256$ nucleotides, a particular sequence of 8 bases is likely to occur once in every $4^8 = 65\,536$ nucleotides and a particular sequence of 16 bases is likely to occur once in every $4^{16} = 4\,294\,967\,296$ nucleotides.

■ Use the information above to estimate how often a particular sequence of 2 bases is likely to occur along a molecule of template DNA.

You will have spotted that a particular sequence of n bases is likely to occur every 4^n nucleotides in a DNA molecule. This means that a primer with only two bases is likely to attach to template DNA every $4^2 = 16$ nucleotides and our PCR would amplify fragments of DNA only 16 nucleotides long. These would be too short to be useful. Using primers that are too short would also make it almost impossible to amplify an entire gene during gene cloning.

■ **Explain why using short primers would interfere with amplifying a gene.**

As we have seen, short primers attach at short distances along a DNA molecule. Therefore, it is highly likely that short primers will attach to a DNA molecule at several points within a gene. As DNA polymerase copies DNA from one primer to the next, only fragments of the gene would be copied, not the whole gene.

By using a variety of primers, molecular biologists could collect an array of DNA fragments. It is much easier for them to find a gene in a DNA fragment than in the entire genome of an organism. It is also easier for them to sequence an organism's genome fragment by fragment. For an increasing number of organisms, molecular biologists have been able to determine the base sequence of genes and of the DNA at either end of genes.

■ **How will knowledge of the base sequence of the DNA at either end of a gene help molecular biologists to amplify only a single gene using the PCR?**

If they know the base sequences at each end of a gene, molecular biologists can design primers that will attach to them. Provided the primers do not attach within the gene, DNA polymerase will amplify the entire gene plus any bases to either side. You learnt in Chapter 10 that genes are controlled by promoter regions that lie close to them. If they wished to transfer a human gene into a bacterium, molecular biologists would need to amplify the gene and its promoter region.

■ **Suggest why molecular biologists might need to amplify a human gene and its promoter region if they wish to transfer the gene successfully into a bacterial cell.**

Bacteria are prokaryotic cells whereas human cells are eukaryotic. Prokaryotic cells have a mechanism for controlling expression of their genes that is different from the mechanism in eukaryotes. Without its promoter region, the human gene might not be expressed in the bacterial cell.

Gene cloning

You learnt in your AS course that a clone is a group of cells, or organisms, that are genetically identical to each other. **Gene cloning** involves making identical copies of a gene. There are two major technologies that enable molecular biologists to clone genes (Figure 10).

We have seen above how, so long as molecular biologists can design primers that mark the beginning and end of a gene, the PCR can be used to amplify a gene. This is one form of gene cloning. Because the PCR occurs in a test tube, this method of gene cloning is called *in vitro* **gene cloning** (literally 'in glassware'). You are probably familiar with the term *in vitro* in

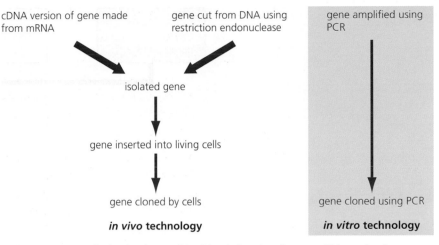

cDNA version of gene made from mRNA

gene cut from DNA using restriction endonuclease

gene amplified using PCR

isolated gene

gene inserted into living cells

gene cloned by cells

in vivo **technology**

gene cloned using PCR

in vitro **technology**

▲ **Figure 10** Gene cloning involves making identical copies of a gene. This can be done using two different technologies: *in vivo* and *in vitro*.

the context of *in vitro* fertilisation (IVF). In an IVF clinic, the fertilisation of an egg cell by a sperm cell occurs in glassware.

An alternative way of cloning a gene involves isolating the gene and inserting it into a suitable host cell. As the host cell divides, it replicates the inserted gene along with its own DNA. Since this method of gene cloning requires the use of living cells, it is called ***in vivo* gene cloning** (*in vivo* means 'in life').

An overview of *in vivo* gene cloning using bacteria

Because they are easy to culture and increase in number at a very fast rate, bacteria are often used to clone genes. This involves the following steps, summarised in Figure 11 on the next page.

- **Step 1.** A fragment of DNA containing one or more genes is obtained by one of the three methods described on page 235.
- **Step 2.** The DNA fragment is inserted into a vector which will transfer it into a bacterium. In this case, the vector is a small DNA molecule called a **plasmid**.
- **Step 3.** The vector is transported into a bacterial cell.
- **Step 4.** The bacterium is allowed to multiply.
- **Step 5.** In reality, only some bacteria will have successfully taken up the plasmid containing the target gene(s). In step 5, these bacteria are identified so that they can be cultured. Any that have not taken up the target gene(s) are destroyed.

Inserting DNA fragments into vectors

Fragments of DNA can be produced by any of the three methods shown in Figure 11. Each method has been described in some detail earlier in this chapter. Having obtained an appropriate DNA fragment, it must be inserted into a **vector** that will carry it into a target cell. Viruses, including those that infect bacteria (bacteriophages) can be used, but plasmids are the most commonly used vectors. **Plasmids** are small, circular DNA molecules that lead an independent existence in many bacterial cells. Each plasmid carries a small number of genes that are expressed in the phenotype of the

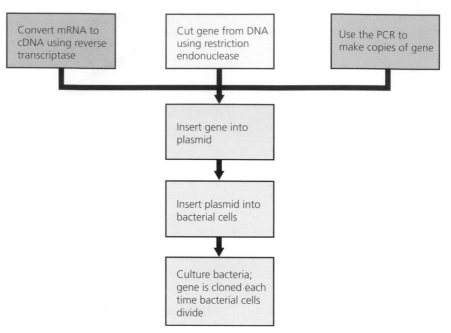

Figure 11 The basic steps in *in vivo* gene cloning. Although other cells can be used, bacteria are the most common host cells for this technology.

Q6

Extracts from bacterial cells will contain the DNA of the bacterium as well as plasmids. Suggest one property of plasmids that will allow them to be separated from bacterial DNA for use later as vectors.

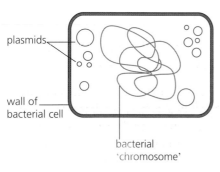

Figure 12 This bacterial cell contains many plasmids. They vary in size but they are smaller than the bacterial 'chromosome'.

bacterium. For example, the ability to survive toxic concentrations of antibiotics, such as ampicillin, chloramphenicol and kanamycin, is often due to the presence inside the bacterium of a plasmid carrying genes for antibiotic resistance. Figure 12 shows that a single bacterial cell can contain many plasmids, each of which is much smaller than the circular DNA that makes up the bacterial 'chromosome'.

Before it can be used as a vector, a plasmid must be cut open. This is done using a **restriction endonuclease**. The DNA fragment can now be inserted into the cut plasmid. Figure 13 shows how, under the right conditions, the ends of the DNA fragment and of the plasmid join together. This process is called **ligation** and is catalysed by a **ligase** enzyme. Some ligases join together DNA fragments and plasmids that have blunt ends. However, ligation is much more efficient if the plasmid and the DNA fragment have sticky ends (page 236). This allows complementary base pairing to hold the two ends of DNA together whilst the ligase joins them up. If either the DNA fragment or the plasmid does not have sticky ends, they can be added before ligation occurs.

If successful, the result of this process is a plasmid that contains a fragment of foreign DNA. Because it contains DNA from two sources, the DNA of this plasmid is referred to as **recombinant DNA**. This gives rise to the scientific name of *in vivo* gene cloning, i.e. **recombinant DNA technology**.

Transferring vectors into host bacteria

Most species of bacteria are able to take up plasmids. There are several methods for getting vectors into bacterial cells. Most were found by trial and error and examiners will not expect you to recall any of them. One of the earliest methods involves soaking bacterial cells, together with plasmids, in an ice-cold solution of calcium chloride followed by a brief

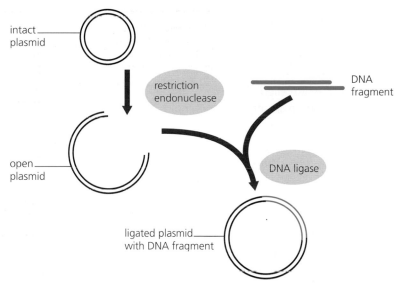

intact plasmid

restriction endonuclease

DNA fragment

open plasmid

DNA ligase

ligated plasmid with DNA fragment

▲ **Figure 13** The cut ends of a plasmid and of a DNA fragment are joined together by a ligase enzyme to form recombinant DNA. This process is much more efficient if the plasmid and DNA fragment have sticky ends rather than blunt ends.

heat shock, during which the temperature is raised to 42 °C for 2 minutes. Quite why this works is poorly understood, but the treatment increases the uptake of plasmids into the bacteria. Bacteria that have taken up plasmids are said to be **transformed** and, because they contain a gene from another organism, they are also **transgenic**.

Identification of transformed bacteria

When presented in a diagram, such as Figure 13, the process for producing transformed bacteria seems simple enough. In reality, biologists cannot see the bacterial cells or the plasmids. They carry out the procedures described above not knowing whether:

- any plasmids have recombinant DNA or whether they have just joined back on themselves (**self-ligated**)
- any bacterium has taken up a plasmid
- any bacterium that has taken up a plasmid has taken up a plasmid containing recombinant DNA. Only about 0.01% of them do.

Biologists need methods to identify which bacteria have been transformed, so that they can grow these, and only these, in a culture medium. There are many ways that they can identify transformed bacteria. Most involve the use of **marker genes** on the plasmids. An examiner can expect you to interpret information given to you about any method of identifying transformed bacteria but cannot expect you to recall any particular method.

Finding transformed bacteria

Escherichia coli bacteria are often used in gene cloning experiments. This bacterium is able to hydrolyse lactose into its constituent monosaccharides.

1 Name the monosaccharides formed by the hydrolysis of lactose.

E. coli uses a series of enzyme-controlled reactions to hydrolyse lactose. One of the enzymes in this series is β-galactosidase. The enzyme is normally encoded by the bacterial gene *lacZ*. The normal *lacZ* gene contains segments that encode different peptide portions of the β-galactosidase molecule. Some strains of *E. coli* have a *lacZ* gene which lacks a segment, called *lacZ'*, that encodes the α-peptide portion of the enzyme.

2 Does β-galactosidase have a quaternary structure? Justify your answer.

Figure 14 a) shows a plasmid, called pUC8, which is commonly used as a vector. This plasmid contains two genes that are of interest to us: one confers resistance to ampicillin and the other is the *lacZ'* gene. If a cell of *E. coli* that lacks the *lacZ'* gene contains a copy of the normal pUC8 plasmid shown in Figure 14 a), it can produce normal β-galactosidase. Figure 14 b) shows where a foreign gene can be inserted into the pUC8 plasmid using a restriction enzyme called *Bam*HI. Note that the recognition sequence of *Bam*HI is in the middle of the *lacZ'* gene in the plasmid.

3 Explain why a cell of *E. coli* that lacks the *lacZ'* gene is able to make β-galactosidase if it also contains a copy of the normal pUC8 plasmid, shown in Figure 14 a).

Having followed the protocol to splice foreign DNA into pUC8 plasmids and then insert the plasmids into cells of *E. coli*, molecular biologists cultured the *E. coli* in a suitable medium.

4 What is a protocol?

After some time, the biologists inoculated the bacteria on to agar plates containing ampicillin and X-gal. Figure 15 shows the appearance of one agar plate after it had been incubated for 48 hours. The circles represent *E. coli* colonies that had grown on the agar. Each colony is a clone of a single *E. coli* cell that was able to grow on the agar.

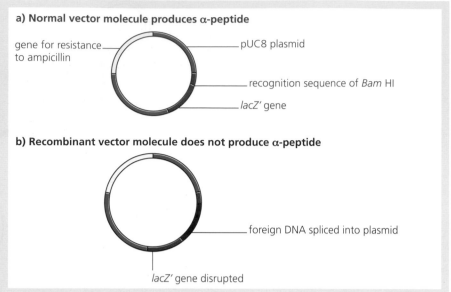

a) Normal vector molecule produces α-peptide

gene for resistance to ampicillin — pUC8 plasmid — recognition sequence of *Bam* HI — *lacZ'* gene

b) Recombinant vector molecule does not produce α-peptide

foreign DNA spliced into plasmid — *lacZ'* gene disrupted

▲ **Figure 14** The plasmid pUC8 is commonly used as a vector in gene cloning experiments. **a)** the normal plasmid with two of its genes indicated; **b)** a recombinant plasmid, containing a foreign gene.

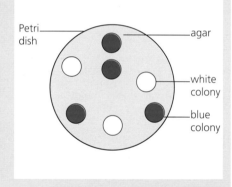

Petri dish — agar — white colony — blue colony

▲ **Figure 15** This agar plate was inoculated with bacteria. The circles represent colonies of bacteria that have grown on the agar plate. The agar contains the antibiotic ampicillin and X-gal. This helps us to identify which bacteria have taken up a transformed pUC8 plasmid.

5 All the colonies in Figure 15 are clones of cells that had taken up the pUC8 plasmid. Explain why this is a valid conclusion.

6 Before inoculating the bacteria on the agar plates, the scientists cultured the bacteria in a suitable medium. Use your knowledge of gene expression to suggest why.

In addition to ampicillin, the agar contained X-gal. This is a lactose analogue. It is white but is broken down by β-galactosidase to produce a blue-coloured product.

7 Use your knowledge of enzyme action to suggest the meaning of 'lactose analogue'.

8 Which of the colonies shown in Figure 15 contain bacteria that can produce β-galactosidase?

9 Use Figure 15 and your answer to Q8 above to explain how we can identify bacteria that have taken up plasmids containing the foreign gene.

Growth of transformed bacteria

Once the transformed bacteria have been identified, they can be removed from the agar plate on which they are growing and transferred to a suitable container for large-scale culturing. They will produce the protein encoded by the foreign gene and this can be extracted. If the protein is secreted by the bacteria, it can be extracted from the medium in which they are grown. If the protein is not secreted by the bacteria, the bacteria must be broken up to allow extraction of the protein. Once extracted, the protein encoded by the foreign gene can be purified and packaged (referred to as downstream processing) before it is sold.

Relative advantages of *in vivo* and *in vitro* gene cloning

We have looked at two methods of cloning genes: the *in vivo* method, using bacteria, and the *in vitro* method, using the PCR.

Each cycle of the PCR takes between 3 and 5 minutes. Normally, 25 to 30 cycles are carried out in any one experiment. Consequently, a PCR experiment takes only a few hours to produce a large gene clone. In contrast, the *in vivo* method takes several weeks. This gives the PCR such a time advantage that you might wonder why scientists continue to use the *in vivo* method at all. Table 1 compares the relative advantages of the PCR and of the *in vivo* method and explains why the *in vivo* method continues to be used.

In vitro method, using the PCR	*In vivo* method, using bacteria
The PCR can copy DNA that has been partly broken down. This makes it useful in forensic science.	Partly broken down DNA is not copied.
The PCR is very sensitive. Even minute amounts of DNA, such as that contained in a single cell, can be copied.	*In vivo* methods are less sensitive, so large amounts of sample DNA are needed. This makes *in vivo* methods less useful in forensic work.
DNA that is embedded in other material can be copied. This makes the PCR useful for analysing DNA in formalin-fixed tissues or in archaeological remains.	Unless DNA can be isolated from the medium in which it is embedded, it cannot be copied by *in vivo* methods.
The cloned genes are produced in solution. They cannot be used directly to manufacture the protein that they encode.	The cloned genes are already inside cells that will manufacture the protein that they encode.
DNA polymerase will only copy a gene if it is marked at each end by complementary primers. If we do not know the base sequences at each end of the gene, we cannot make appropriate primers. Thus, we cannot use the PCR to copy genes that have not been studied before.	Once incorporated into its host DNA, the gene will be copied by the host cell. This method can be used to copy genes that have not been studied before.
The PCR becomes unreliable when we use it to copy DNA fragments longer than about 1000 base pairs (1 kbp).	*In vivo* methods reliably copy genes up to about 2 Mbp long.
In vitro methods lack error-correcting mechanisms, so the error rate is higher than in cell-based methods.	Cells have mechanisms for correcting any errors that are made when copying genes. This reduces the error rate of gene copying.

Table 1 The relative advantages of the PCR and cell-based methods of cloning genes.

Chapter 12
Making use of gene cloning and DNA analysis

To stay healthy, we must have certain nutrients in our diets. One of these is vitamin A, contained in eggs, oily fish, meat, and milk. People who do not eat these foods lack vitamin A in their diet. However, fruit and vegetables contain β-carotene, which their bodies can convert into vitamin A. Like the children shown in Figure 1, millions of people worldwide live on a diet of white rice (rice which, unlike brown rice, has had its outer husks removed). White rice contains neither vitamin A nor β-carotene. As a result, vitamin A deficiency is common in people living on a rice diet. To put this into context, the World Health Organization (WHO) estimates that over 100 million children suffer vitamin A deficiency, resulting in about half a million children going blind every year.

▲ **Figure 1** Families in underdeveloped countries subsist on a diet consisting largely of polished white rice. Although it contains carbohydrates, which are an important source of energy, polished rice contains no vitamin A or β-carotene.

'Golden Rice' was first developed in 2000 as a solution to this problem. Genes encoding the many enzymes needed to make β-carotene were transferred into rice plants so that their endosperm, the shiny white part that we eat as white rice, contained β-carotene. Once absorbed from their diet, humans can convert β-carotene into vitamin A, so avoiding the deficiency symptoms. The scientists who developed 'Golden Rice' intended to give the rice seeds to farmers for free and, since it was seen as for the public good, their research work was publicly sponsored. Despite this, there were many critics of their work. You can read about some of this criticism, and about the work that the team of scientists followed up to address some of the scientific issues, on www.goldenrice.org.

Transgenic organisms can be useful to humans

A **transgenic** organism is one that contains a foreign gene. You saw in Chapter 11 how genes can be transferred into bacterial cells, producing transgenic bacteria. When a gene is successfully transferred into another organism, the transgenic organism will produce the protein encoded by the transferred gene.

Collecting human proteins from transgenic organisms

Many human diseases result from an inability to produce a vital protein. For example, Type I diabetes (page 208) is caused by failure to produce insulin and haemophilia (page 115) is caused by failure to produce an essential blood-clotting protein. If the gene encoding one of these proteins is successfully transferred into a transgenic organism, the protein it produces can be extracted and used to treat patients with the relevant disease. Table 1 on the next page shows a range of organisms that have been genetically modified to produce human proteins on a commercial scale.

Extraction of the human protein is much easier if the transgenic organism secretes it into the fermentation medium in which it is growing (Figure 2, overleaf) or in its milk. This explains the presence of microorganisms and mammals listed in Table 1. Plants do not secrete large volumes of protein; extraction of a human protein from plants is easier if we know where in the plant it is produced. Scientists can ensure that a human protein is secreted in milk or deposited only in the seeds of plants by transferring the human gene into a specific part of the genome of the transgenic organism. By placing it downstream of a promoter controlling lactation in mammals, or of a promoter controlling seed production in plants, scientists control where in the transgenic organism the human protein is made. Producing pharmaceutical products using transgenic farm animals (sometimes called 'pharming') and transgenic plants is an increasingly important source of new drugs.

Q1

Unlike eukaryotic cells, bacterial DNA does not contain introns.
a) Suggest the advantage of transforming bacteria by inserting complementary DNA (cDNA) rather than a copy of the original gene.
b) Suggest one advantage of using transgenic yeast rather than transgenic bacteria to produce human proteins.

Transgenic organism	Human protein	Clinical use of protein
Bacteria	Alpha-1-antitrypsin	Treatment of emphysema
	Growth hormone	Treatment of dwarfism
	Insulin	Treatment of Type 1 diabetes
Yeast	Alpha-1-antitrypsin	Treatment of emphysema
Plants	Collagen	Reconstructive surgery
	Growth hormone	Treatment of dwarfism
	Interleukins	Treatment of cancer
	Vaccine	Prevention of measles
Cattle	Fibrinogen	Treatment of blood clotting disorders
Goats	Anti-thrombin	Treatment of deep-vein thrombosis
Pigs	Factor VIII	Treatment of haemophilia

Table 1 The range of transgenic organisms that have been used in the commercial production of human proteins. Many more are at the pre-clinical trial stage of production.

▲ **Figure 2** These large industrial fermenters contain a culture of transgenic bacteria that produce a human protein. Since the bacteria secrete the human protein, it is relatively simple to extract it from the culture medium.

▲ **Figure 3** The damage to this plant is called a gall. It is caused by a bacterium, *Agrobacterium tumefaciens*, which has infected the plant. When *Agrobacterium tumefaciens* infects a plant cell, it injects its own plasmid, a T-plasmid, which the plant cell incorporates into its own DNA. By inserting genes into T-plasmids, biologists have been able to transfer foreign genes into plants.

Transgenic plants produce genetically modified (GM) food

In addition to producing human proteins, research into gene transfer in plants has focused on improving plant productivity. In particular, research has focused on two plant characteristics: resistance to insect pests and resistance to herbicides. One species of soil bacterium, *Agrobacterium tumefaciens*, has been especially important in transferring new genes into crop plants. Figure 3 explains why this bacterium has been so important.

Agrobacterium tumefaciens produces chemicals that make the bacterium resistant to attack by insects. The genes encoding the production of these **natural insecticides** have been successfully transferred from bacteria to crop plants, reducing the need to spray crops with insecticides. In 1995, maize was the first plant to be marketed which had been genetically modified in this way, with potato plants and cotton plants soon following them on to the market. Today, there are many competing agrochemical companies around the world, marketing many types of transgenic insecticide-producing crop plants.

No one has yet produced a plant that is resistant to weeds growing around it. Instead, transgenic plants have been produced that are resistant to the most commonly used herbicide, **glyphosate** (sold commercially as 'Roundup'). Glyphosate inhibits an enzyme (EPSP synthase) involved in the pathway which plants use to make essential amino acids. Without a functional EPSP synthase, weed plants cannot make some of the proteins they need and die. Unfortunately, the same happens to crop plants that have been sprayed to get rid of weeds. The solution was to transfer into crop plants a bacterial gene that confers resistance to glyphosate. Such herbicide-resistant crop plants (called 'Roundup Ready') are routinely grown in many countries.

The food produced by transgenic plants is commonly called genetically modified food (**GM food**). It was hoped that transgenic plants would increase crop production and so help to reduce starvation in the world. You would probably support this aim, to reduce world starvation, but are you in favour of genetically modified food? The issue has caused great debate in the UK. The growth of transgenic crops in the UK is restricted by government legislation and any food produced by transgenic plants must be clearly labelled as GM food. Many people refuse to buy GM food and are angry when they find that the constituents of processed food include GM products. Some people, like the activists in Figure 4, take extreme action and destroy experimental GM crops where they discover them.

People's concerns about GM foods tend to fall into one of three categories.

- **Does eating GM foods harm us?** There are strict rules in the UK about trialling new medicines before they can be used on humans. Some people ask why there are not similarly strict rules about using GM food in the human food chain.
- **Do GM plants adversely affect the environment?** Many people are concerned about the damage that pesticides might do to natural communities. Some people point out that the use of herbicides seems to

▲ **Figure 4** Some people take extreme measures in support of their opinions. These activists are destroying what they believe to be a crop of transgenic plants (GM plants).

Q2

In addition to the DNA found in plant chromosomes, chloroplasts contain their own DNA. Genes conferring resistance to insecticides and herbicides are commonly inserted into the DNA of chloroplasts, rather than into the transgenic plant's chromosomes. Suggest why this technique reduces the risk of horizontal transfer of foreign genes by pollen grains.

have increased where herbicide-resistant GM crops have been grown. Others are concerned that **horizontal gene transfer** between plant species might result in plants other than the GM crop gaining the transferred genes for herbicide resistance.

- **Does globalisation disrupt local enterprise?** These concerns are about the dominance of large, multinational companies and the effect of their behaviour on small farmers, rather than about the science of genetic modification. The seeds of genetically modified plants are produced by a small number of multinational companies. They sell these seeds as part of a package; the farmer must agree to buy seeds, fertiliser and pesticides from the same company. Tying in to a single provider increases the risk to the farmers of changes in company policy or of changes in political relations between countries.

You might like to discuss the GM issue in class, including why people in some countries appear to be less concerned about GM crops and GM foods than people in the UK.

Gene therapy is used to supplement defective genes

Many people object to the concept of producing transgenic humans, yet this is what gene therapy does. The aim of gene therapy is to treat an inherited disease by providing the sufferer with a corrected copy of their defective gene. The process of putting a corrected gene into a cell is called **transfection** and the cell that has received the new gene is said to have been **transfected**. There are two approaches to gene therapy: **somatic cell therapy** and **germ cell therapy**.

Somatic cell therapy

During **somatic cell therapy**, copies of a corrected gene are inserted directly into the body cells of sufferers. In some cases, body cells are removed, copies of the corrected gene inserted into them and the transfected cells put back in the same patient's body. This works well with blood diseases, such as leukaemia, where cells from the patient's bone marrow are extracted, transfected and replaced. In other cases, where large numbers of body cells cannot be removed safely, the corrected genes are inserted directly into the affected tissues. This technique has been used to treat lung diseases, such as cystic fibrosis.

Vectors are needed to transfer corrected genes into the cells of a sufferer. **Retroviruses** are often used as vectors for bone marrow because they have been found to transfect a large number of stem cells successfully. Viruses, known as adenoviruses, have also been used as vectors for treating cystic fibrosis, as have **liposomes** – small droplets of lipid that fuse with the surface membranes of epithelial cells in the lungs (Figure 5).

When stem cells from bone marrow are transfected and replaced in the marrow of patients, all the different types of white blood cells produced from them contain the added gene. This provides a long-term treatment for blood diseases, such as leukaemia. Where stem cells cannot be used, the effects of the corrected gene last only as long as the transfected cells remain alive. This explains why gene therapy to treat cystic fibrosis must be repeated every few weeks.

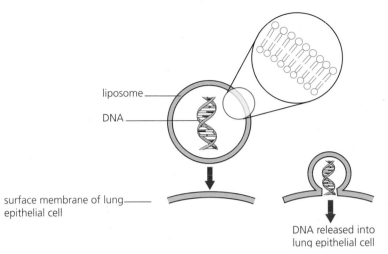

liposome

DNA

surface membrane of lung epithelial cell

DNA released into lung epithelial cell

▲ **Figure 5** The corrected gene for cystic fibrosis can be carried into the epithelial cells of the lungs by small droplets of lipid, called liposomes. A liposome is able to fuse with the surface membrane of a lung epithelial cell, releasing its DNA into the cell.

Q3

Somatic cell therapy is more successful in treating diseases caused by a recessive allele of a gene than those caused by a dominant allele. Suggest why.

Q4

Some people are concerned that the techniques of germ cell therapy could be misused. Suggest one way in which the techniques could be misused.

Q5

How could the techniques of germ cell therapy be used to produce cattle that secrete a human protein in their milk?

Attempts have been made to use gene therapy to treat cancer. As you know from Chapter 10, cancers can result from activation of **oncogenes** or from inactivation of tumour suppresser genes. Gene therapy has been used to insert corrected copies of active tumour suppresser genes into cancer cells and to attempt to prevent the expression of an oncogene in cancer cells. The success of this research is currently hampered by a lack of vectors that will ensure the corrected genes are taken up by cancer cells.

Germ cell therapy

In **germ cell therapy**, a corrected gene is inserted into an egg cell that has been fertilised using *in vitro* fertilisation techniques (IVF). Once fertilised, the embryo is allowed to develop before being re-implanted into the mother's uterus. If gene transfer is successful, mitosis will ensure that every cell in the embryo contains a copy of the corrected gene. As an adult, the transfected individual will pass on copies of the corrected gene to her or his offspring. Consequently, germ cell therapy provides a treatment for an inherited condition that not only lasts the lifetime of the sufferer but also crosses generations.

Medical screening and genetic counselling

Many human diseases result from mutations of genes, which give rise to alleles that are useful in one context but not in another. You learnt about one such example in Chapter 5, where sickle-cell anaemia results from an allele of the gene encoding β-globulin, one of the two types of polypeptide in a haemoglobin molecule. In areas of the world where malaria is endemic, heterozygotes for this gene are at a selective advantage because they are resistant to malaria. In countries where there is currently no malaria, for example the UK, heterozygotes are at a disadvantage because some of their children might be homozygous for the sickle-cell allele. These children will suffer uncomfortable and life-threatening symptoms.

A couple intending to have children might be concerned if they are aware that there is an inherited disease, such as sickle-cell anaemia, in their family. They might take medical advice about the risk of having a baby with the inherited condition. This advice is called **genetic counselling**. In the past, genetic counsellors were able to use the sort of information in Chapter 4 to work out the chances of an affected baby being born. For example, you can use your understanding of monohybrid inheritance to work out that the chances of two people who are both heterozygous for the sickle-cell allele having a baby with sickle-cell anaemia is 1 in 4 (a probability of 0.25).

Breast cancer and ovarian cancer

Each year in the UK about 200 000 cases of breast cancer and about 25 000 cases of ovarian cancer are diagnosed. Research has shown that about 10% of these cases are inherited. At least two genes are involved in the development of breast cancer: BRCA 1 (**BR**east **CA**ncer 1) and BRCA 2. Normally, the BRCA 1 and BRCA 2 genes control cell growth but mutations in these genes increase the risk of breast and ovarian cancers.

1 What type of gene is BRCA 1 and BRCA 2?

We each have two copies of the BRCA 1 and BRCA 2 genes, inheriting one from each parent. Someone who carries one of the mutant alleles of either gene has an increased risk of developing breast cancer.

2 What does this tell you about the nature of the mutant alleles of the BRCA 1 and BRCA 2 genes?

Table 2 shows data about the risk of a woman developing breast cancer and ovarian cancer in the UK.

3 The risk of developing cancer in the middle column of Table 2 is shown as a range, whereas the risk in the right-hand column is shown as a single value. Suggest why.

4 Give two conclusions about the effect of carrying a mutant allele of a BRCA gene on the risk of developing breast cancer.

5 Suggest why the risk of developing breast cancer is higher than the risk of developing ovarian cancer.

Risk of developing:	Carrier of one mutant allele of BRCA 1 or BRCA 2 gene	General population
breast cancer by age 50 / %	33–50	2
breast cancer by age 70 / %	56–87	7
ovarian cancer by age 70 / %	27–44	< 2

Table 2 The risk of developing cancer among women who are carriers of one mutant allele of the BRCA 1 gene or BRCA 2 gene compared with that of the general population.

Genetic counsellors do not only advise people about their chances of having a baby with an inherited disease or about their chances of developing an inherited disease, such as cancer, later in life. They also provide emotional and medical support to the affected families. This is needed because the results of screening tests provoke difficult decisions. For example, a woman who finds that she is carrying a child with an inherited disease might have to make a decision about carrying her pregnancy to full term or having an abortion. Similarly, a woman who finds she has an increased risk of developing breast cancer might have to decide to have her healthy breasts surgically removed so they cannot later develop cancer.

None of these decisions is easy to make and affects not only the person making the decision but other members of their families as well.

It is likely that shortly after you were born, a drop of blood was taken from your heel. This 'pin prick' procedure collected blood which could be used to test for the presence of common inherited disorders, such as phenylketonuria. This represents a simple form of medical screening. Advances in DNA technology have resulted in much more sophisticated medical screening procedures. Using the techniques you learnt about in Chapter 11, it is now possible to analyse DNA taken from scrapings of cells from inside your cheek, cells taken from the amniotic fluid surrounding an embryo in its mother's uterus or from the umbilical cord connecting the embryo to its placenta. **DNA hybridisation** and **DNA sequencing** are two important techniques used in DNA analysis. We need to look at how these techniques work.

Using DNA hybridisation to find harmful genes

If we know at least part of the base sequence of a harmful gene, we can use **DNA hybridisation** to find out if this gene is present in DNA samples. Before hybridisation, the test DNA is treated using techniques you learnt about in Chapter 11. Let's revise these techniques.

First, the DNA is extracted from the sample of cells taken from a patient. After purification, the test DNA is amplified using the PCR.

■ Why is the test DNA amplified?

The test sample will only contain a few cells, for example cheek cells, cells from the amniotic fluid or from the umbilical cord. These cells only yield a small amount of DNA. We need a large amount of DNA for analysis, so the PCR is used to produce this extra DNA. Of course, the PCR produces DNA that is identical to the test DNA.

The amplified test DNA is then digested, using a restriction endonuclease.

■ Suggest why the DNA is digested.

Whole DNA molecules are too long to be analysed successfully in one go. This is why the test DNA is digested using restriction endonuclease.

The restriction fragments are then separated using gel electrophoresis.

■ Explain why gel electrophoresis is able to separate DNA restriction fragments.

Since DNA has a negative charge, DNA fragments migrate to the positive electrode of an electric field. The gel through which they move has fine pores. Small fragments of DNA move faster through these pores than large fragments. As a result, the fragments are separated, with the smallest nearer the positive electrode, forming bands in the gel.

■ After separation, the bands of restriction fragments are transferred to a nylon membrane, using the Southern transfer technique. Explain the advantage of this transfer.

Because the gel used in electrophoresis is fragile, it is likely to break if we handle it. Transferring the DNA bands to a strong nylon membrane gives us a more robust record of the restriction fragments.

In the final preparatory stage, the DNA fragments on the nylon membrane are treated to break the hydrogen bonds holding the DNA together.

■ Describe the structure of the restriction fragments held on the nylon membrane after treatment to break their hydrogen bonds.

Hydrogen bonds between complementary base pairs hold together the two polynucleotide chains in a DNA molecule. When these bonds are broken, the nylon membrane contains bands of single-stranded DNA fragments. These fragments anneal with any strands of nucleotides that have complementary base sequences. This is where labelled DNA probes come in.

At least part of the base sequence of the harmful gene is known. This enables us to produce single-stranded DNA probes with a complementary sequence. These probes will anneal to any complementary DNA fragment on the nylon film. This is the process of **DNA hybridisation**. If the DNA probe is labelled, its position can be found. Figure 6 shows how a DNA probe will attach to a complementary base sequence if it is present in the test DNA and how, if labelled, we can find the hybridised DNA after electrophoresis of the restriction fragments.

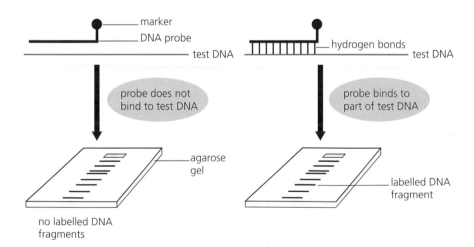

▲ **Figure 6** A labelled DNA probe can be made which has a sequence that is complementary to part of the harmful allele but not to the normal allele of a gene. Once the DNA probe has attached to a fragment of the harmful allele, we can find it in the separated restriction fragments.

■ Use Figure 6 to explain how DNA hybridisation enables biologists to locate the harmful allele of a gene in a sample of DNA taken from, for example, cells in amniotic fluid.

Since the restriction fragments were made from the entire DNA in the sampled cells, one of them will contain all, or part, of the gene in question, for example the gene for β-globulin. The sequence of bases that is different in the harmful allele is known. The labelled DNA probe has a base sequence that is complementary to that part of the harmful allele. If the label attaches to any of the bands of restriction fragments on the nylon membrane, the DNA fragment in that position must be from the harmful allele. If no label attaches to any band, the DNA does not contain the harmful allele of the gene.

How chain termination sequencing finds the base sequence of an unknown gene

DNA hybridisation is only useful if we know at least part of the base sequence of the gene we are looking for. **DNA sequencing** finds the base sequence of genes that have not been studied before. The methods for doing this are all based on the use of **dideoxyribonucleotides**. Look at Figure 7, which shows molecules that you might recognise. Molecule a) is ribose, the 5-carbon sugar found in every RNA nucleotide. Molecule b) is deoxyribose, the 5-carbon sugar found in every DNA nucleotide. Note that it has one oxygen atom less than ribose – hence the name *deoxyribose* (page 215). Now look at molecule c). Can you see how this differs from ribose and deoxyribose? It has one less oxygen atom than deoxyribose and two less than ribose: this is *di*deoxyribose.

Just like ribose and deoxyribose, a molecule of dideoxyribose can join up with phosphate and an organic base to form a nucleotide. During DNA replication, a nucleotide containing dideoxyribose (a **dideoxynucleotide**) can pair with a deoxynucleotide on the template strand that has a complementary base. However, DNA polymerase stops replicating DNA when it encounters a nucleotide containing dideoxyribose on the developing strand. This is the basis of the **chain termination technique** for sequencing DNA.

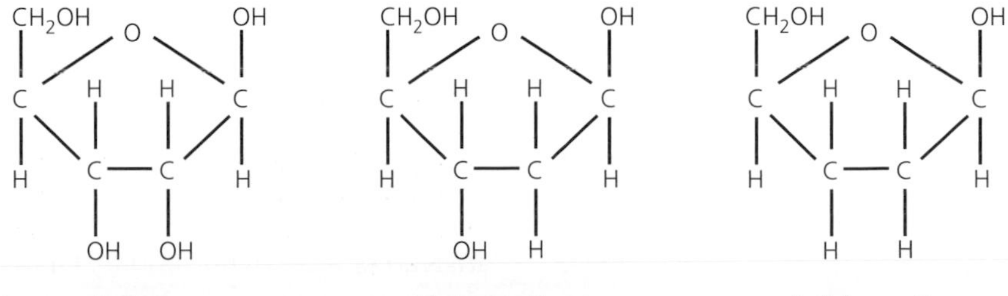

a) ribose   b) deoxyribose   c) dideoxyribose

▲ **Figure 7** A molecule of **a)** ribose, **b)** deoxyribose and **c)** dideoxyribose.

Automated DNA sequencing

Figure 8 a) shows a single-stranded version of the DNA we wish to sequence. It has been cloned into a vector, which in this technique is usually the DNA of a bacteriophage called M13. A short oligonucleotide has been annealed to the DNA to act as a primer.

DNA polymerase is added to the recombinant DNA, together with a mixture of deoxynucleotides containing each of the four bases: adenine, cytosine, guanine and thymine. Under normal circumstances, the DNA polymerase attaches to the primer and begins the process of replicating the recombinant DNA. The new strand starts to grow as free nucleotides form hydrogen bonds with complementary bases in the starting strand (Figure 8 b).

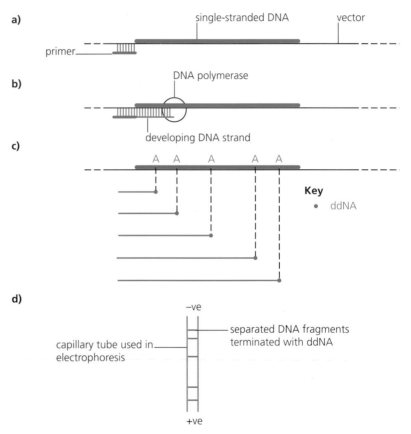

▲ **Figure 8** The principles involved in the use of an automated DNA sequencer. **a)** The length of single-stranded DNA to be sequenced is cloned into a vector and a primer added. **b)** DNA polymerase attaches to the primer and begins DNA replication, adding nucleotides with bases that are complementary to the bases on the recombinant DNA. **c)** The DNA polymerase inserts a dideoxynucleotide and replication of the chain terminates. **d)** Provided the ratio of normal (deoxynucleotides) to dideoxynucleotides is high enough, chains of one to several hundred nucleotides are produced.

However, also present in the mixture of nucleotides are some dideoxynucleotides. When, by chance, DNA polymerase inserts one of these instead of a deoxynucleotide, DNA replication stops. If the ratio of deoxynucleotides to dideoxynucleotides is high enough, we end up with complementary chains of DNA varying in length from a single nucleotide to several hundred nucleotides. Figure 8 c) shows how this is possible, using just one of the four types of dideoxynucleotides (ddNA). Whenever ddNA forms a complementary base pair with thymine on the recombinant DNA strand, replication stops.

Each type of dideoxynucleotide is labelled with a different fluorescent dye. The dideoxynucleotides carrying adenine (ddNA) are labelled with a green fluorescent dye, ddNC with a blue dye, ddNG with a yellow dye and ddNT with a red dye.

At the end of the incubation period, the newly formed DNA fragments are detached from the template DNA and these new single-stranded chains are separated by size, using a special type of electrophoresis that takes place inside a capillary tube (Figure 8d). The resolution of this technique is so high that it separates strands that are different in length by a single nucleotide.

When illuminated by a laser beam, each of the four dideoxynucleotides fluoresces. The colour and position of the fluorescence is read by a detector which then feeds this information into a computer, which either stores the data for future analysis or produces a printout, like the one in Figure 9. An automated sequencer can read almost 100 different DNA sequences in a period of two hours.

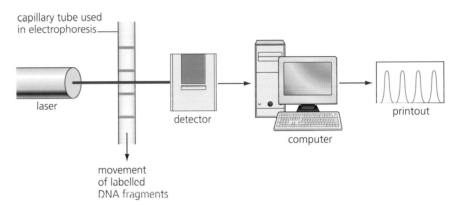

▲ **Figure 9** As the replicated fragments of DNA are separated, they pass through a laser and fluoresce. The fluorescence detector 'reads' the fluorescence and sends the data to a computer for analysis. The coloured printout shows the base sequence of the DNA sample.

Interpreting the results of manual DNA sequencing

The manual method is similar to the automated method above, except that:

- a separate run is made for each type of dideoxynucleotide – ddNA, ddNC, ddNG and ddNT
- after incubation, the four mixtures are placed in separate wells of a gel and separated by gel electrophoresis
- the dideoxynucleotides are labelled with radioactivity, rather than with fluorescent dyes
- a Southern transfer of the gel is made and an autoradiograph made from it.

Reading the DNA sequence is relatively easy. Figure 10 shows part of an autoradiograph. For each dideoxynucleotide base, it shows bands which are DNA fragments where replication was terminated. As you know, the band that has moved the furthest is the smallest fragment. This will be a fragment where replication stopped at the first base.

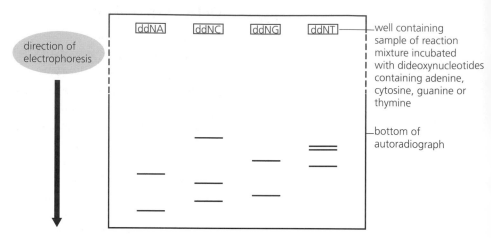

▲ **Figure 10** The wells at the top of this autoradiograph contained a sample of a reaction mixture that had been incubated with dideoxynucleotides containing adenine (ddNA), cytosine (ddNC), guanine (ddNG) or thymine (ddNT), respectively. Below each well is the track of the separated DNA fragments produced by DNA replication in the presence of each dideoxynucleotide. The DNA sequence of the newly synthesised DNA is read by identifying the distance each fragment has moved, starting with the one that has moved furthest.

■ **Look at Figure 10. Which of the bands has moved the furthest?**

The fragment that has moved furthest is in the ddNA track. This is the smallest fragment that was produced when a dideoxyribonucleotide stopped DNA polymerase forming a new strand of DNA.

■ **What is the very smallest fragment of DNA that could be formed?**

The smallest fragment is a single nucleotide. By chance, a dideoxynucleotide was the first to be inserted into the new DNA strand and its presence stopped DNA polymerase inserting any more nucleotides. We have now found the first nucleotide that is incorporated into the new DNA strand. It carries adenine.

■ **Reading up the autoradiograph, what base is carried by the next smallest fragment of DNA?**

The next smallest fragment is in the ddNC track. This is a fragment of DNA that is only two nucleotides long and it carries cytosine. The first two bases in the test DNA are adenine and cytosine.

■ **What is the third base in the DNA sequence?**

The fragment which has moved the third furthest is in the ddNG track. This fragment stopped DNA replication because it carries the ddG nucleotide. So our third base is guanine and the base sequence of the test DNA becomes ACG.

■ **Continue to read the autoradiograph in Figure 10. What is the DNA sequence of the DNA?**

By repeating what we have done so far, you should find that the base sequence is ACGCATGTTC.

DNA fingerprints

You have learnt that any one of our chromosomes carries the same genes in the same order. Consequently, you might expect that we would get the same pattern of restriction fragments if we used the same restriction endonucleases to digest copies of the same chromosome from two different people. In fact, we would not; we would get different patterns of restriction fragments. The reason for this is that, in the non-coding regions of our DNA, each of us has different repeated sequences of bases. These repeated base sequences include **restriction fragment length polymorphisms** (RFLPs, pronounced 'rifflips') and **short tandem repeats** (STRs). Because the length of these repeated base sequences is different in different people, we can use them to identify the source of DNA in tissue samples. This is the basis of **DNA fingerprinting**.

Figure 11 shows how the inclusion of different lengths of repeated base sequences affects the distance between the recognition sites of a restriction endonuclease on two samples of DNA. As a result of these differences, DNA fragments of different lengths would be produced by digestion of these two DNA molecules with the restriction endonuclease. Figure 12 shows how these different lengths of DNA would produce different patterns of restriction fragments after electrophoresis. The important idea to grasp is that only identical twins have the same pattern of non-coding, repeated base sequences and so only they would produce the same pattern of restriction fragments.

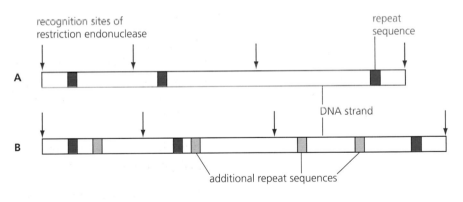

▲ **Figure 11** Each band represents part of a DNA molecule that has been taken from the same chromosome location from two different people, **A** and **B**. The recognition sites of an endonuclease are shown on each DNA molecule. The recognition sites are further apart in **B** because the non-coding regions of this DNA contain more repeated base sequences.

lane **A**   lane **B**

autoradiograph

▲ **Figure 12** The banding patterns produced after the two sets of DNA shown in Figure 11 have been digested using the same restriction endonuclease. Lane **B** contains longer DNA fragments than lane **A** because the DNA of **B** contained more repeated base sequences than the DNA of **A**.

DNA fingerprinting by hybridisation of RFLPs with labelled probes

This is the classical method for genetic fingerprinting. You have come across the techniques involved earlier in this chapter. A sample of DNA is digested into restriction fragments using a restriction endonuclease. The restriction fragments are of different lengths, depending on the number of repeat sequences within them, and are unique to each individual. These fragments are then separated using gel electrophoresis and a Southern blot is prepared from it. Radioactively labelled probes with the complementary base sequence to the repeat sequence are added to the blot. The probes reveal a labelled band wherever they hybridise with a restriction fragment containing the repeat sequences. Thus, each band in the fingerprint represents a DNA fragment that contains the repeat sequence. Figure 12 shows the genetic fingerprints of two people. The pattern of bands is different because the location of the repeat sequences was different in the two sets of DNA.

The very first RFLP to be used in classical DNA fingerprinting had the base sequence GGGCAGGABG, where B represents any base. This RFLP was repeated a different number of times in the non-coding regions of the DNA samples used.

Figure 13 shows the results of the very first RFLP analysis which led to a criminal receiving the death sentence in the USA. A young couple had been murdered whilst they slept in their car. Their bodies were discovered the next day. A post mortem showed they had both died of gunshot wounds and that the woman had been raped. One man was later arrested driving the couple's stolen car. Under police questioning, he identified a friend who had been with him on the night of the murders. DNA from the semen found in the woman's body and a DNA sample from each suspect was digested using the same restriction enzyme, and the restriction fragments separated using gel electrophoresis. If you look at Figure 13, you will see that the DNA fingerprint of the semen does not have the same pattern of restriction fragments as suspect 1 but does have the same pattern as suspect 2. On the basis of this evidence, suspect 2 was found guilty of rape and murder and given a double death sentence. The jury at the time was told that the chance of an innocent person showing the same pattern of restriction fragments was about 1 in 9 billion. At the time of the trial, the human population of the world was less than 6 billion.

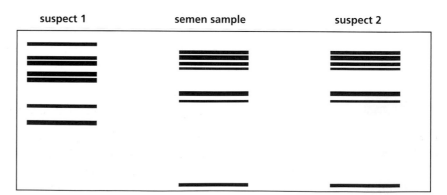

▲ **Figure 13** The results of the first RFLP analysis which led to a criminal receiving the death sentence. The DNA fingerprint of suspect 2 and of the semen taken from a murdered woman show that he was the rapist. He was convicted of rape and murder and received a double death sentence.

The Bains family

Mr Bains first settled in England in 1990. Two years later, he applied to bring his wife and four children to join him. The immigration authorities needed to be sure that the children were entitled to come into England and so they carried out a RFLP analysis on the whole family. Figure 14 shows the results.

Mr Bains had been married before. Sadly, his first wife had died giving birth to their son. Some years later, Mr Bains remarried and he and his new wife had two daughters. They also adopted another son.

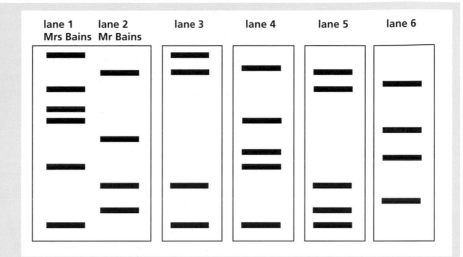

▲ **Figure 14** The DNA fingerprints of the Bains family members.

1 Look at lanes 1 and 2 in Figure 14. They show the separated restriction fragments of Mr and Mrs Bains. Explain why the two patterns of restriction fragments are different.

2 Lane 6 of Figure 14 shows the restriction fragments from Mr and Mrs Bains' adopted son. Explain why this conclusion is valid.

3 Use Figure 14 to identify Mrs Bains' stepson. Explain your answer.

DNA profiling by PCR of short tandem repeats (STRs)

The classical DNA fingerprinting method described earlier has a number of limitations. For example:

- it requires a fairly large sample of DNA so the amount of DNA found in a hair or a bloodstain is often not enough for this technique to be used
- problems with interpreting the bands have also led to DNA fingerprints being insufficient evidence to secure a court conviction.

Consequently, a more powerful technique has replaced the classical fingerprinting technique. This involves DNA profiling.

In addition to the RFLPs, the non-coding regions of our DNA contain **short tandem repeats (STRs)**, also called **microsatellites**. These STRs are repeated blocks of base sequences that are much shorter than the RFLPs. In our own genome, the most common short tandem repeat is a sequence of two bases, cytosine and adenine, repeated between 5 and 20 times, for example CACACACACACACA. The number of repeats in any one STR is variable; in a single population, there might be as many as ten different versions of one STR. In DNA profiling, the different versions of a selected number of STRs are determined.

Using primers that anneal to the DNA to each side of a particular repeat sequence, the PCR is used to make many copies of the DNA containing these STRs. After amplification by the PCR, the DNA fragments are

*Q*₆

Use the information about genetic fingerprinting here and on page 261 to give one difference between a restriction fragment length polymorphism (RFLP) and a short tandem repeat (STR).

separated using gel electrophoresis. Figure 15 shows a gel after DNA profiling. Lanes 2 and 3 show the results of profiling the same STR in two different people. Each person has two different bands for this STR, representing a difference in the repeat sequences in the gene inherited from each parent. Although they have one band in common, these two people have different DNA profiles. Normally, DNA profiling is carried out using several selected STRs, so that many more bands are present in the gel. Lane 4 in Figure 15 shows a run analysing three different STRs in one person.

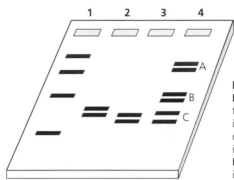

Lane 1: DNA size markers
Lanes 2 and **3:** PCRs of a single short tandem repeat in one gene from two individuals. The two individuals have different profiles but one band (allele) in common
Lane 4: A PCR of short tandem repeats in three genes (labelled A, B and C)

▲ **Figure 15** DNA profiles using STRs.

Just as with the automated sequencing technique earlier, the STR fragments can be labelled with fluorescent dyes and separated using capillary electrophoresis. Scanners 'read' the sequences and feed the data into a computer which compares them with sample material. This technique is much more sensitive than the standard test and is now commonly used by teams of forensic scientists. Figure 16 shows the type of printout from an automated scanner.

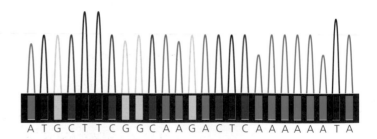

A T G C T T C G G C A A G A C T C A A A A A A T A

▲ **Figure 16** The printout from an automated DNA profiler. You might have seen this type of DNA fingerprint in a film or TV programme about crime scene investigators.

Moral and ethical issues

As a biology student, you should be able to recognise and discuss some of the moral and ethical issues raised by DNA technology. It would be wrong for this book to suggest that there is a 'correct' response. You must make up your own mind.

However, you will be expected to discuss the controversial issues raised by gene cloning and DNA technology using appropriate scientific terminology and understanding. Three examples of how you could show your biological knowledge and understanding are given below.

You might express concern that producing transgenic animals causes the animals to suffer.

* As a biologist, you should be able distinguish between the effects of a foreign gene within an animal and the procedures involved in producing transgenic animals. Whilst the presence of a foreign gene does not appear to cause suffering, the manipulations involved in producing a transgenic animal might cause suffering. For example, animals produced by somatic cell nuclear transfer (page 214) suffer a relatively high rate of birth defects and even the healthy animals appear to suffer from premature aging.
* You might use Dolly the sheep as an example of premature aging. If so, you should make it clear that she was *not* a transgenic animal. Her premature aging was associated with the somatic cell nuclear transfer technique rather than any foreign gene.

You might express concern about the use of genetically modified plants to improve crop productivity.

* As a biologist, you should not justify this concern using statements such as, 'this harms the environment'. You can show knowledge and understanding of biological concepts and principles by giving a specific example of how a GM crop could cause damage. For example, you could say that horizontal gene transfer between bacteria could result in the foreign gene for herbicide resistance being transferred from the GM crop plants to weed plants and that this would cause the weeds to become an even greater pest. You might also recall from page 252 that foreign genes are often inserted into the DNA within chloroplasts to reduce the likelihood of their horizontal transfer to other plants via pollen.

You might express concern about the use of gene therapy to cure human diseases. There is no easy answer to problems such as this, but you can give a balanced argument that includes biological knowledge. For example, you might argue that:

* it is unreasonable to object to the management of cystic fibrosis using liposomes in respiratory inhalers to insert corrected versions of the gene into the lung cells of sufferers
* if bone marrow transplants are acceptable, it is difficult to reject the correction of blood disorders by inserting corrected genes into blood stem cells
* germ cell therapy has been beneficial in producing cows that yield milk with less fat and more protein. However, since this type of genetic manipulation involves changing the genotype of this animal and of future generations of this animal in a directed way, it would be unacceptable to use it with humans.

It is *never* a good idea to use the expression 'playing at God', since this shows no biological knowledge or understanding. Remember that a good scientist uses evidence from reliable sources, and not unsubstantiated opinions, when formulating an argument.

Index

Index

control of blood glucose concentration 204–9
glucose tolerance test 209
glycerate 3-phosphate (GP) 46–7
glycogenesis 206
glycolysis 56, 184–5
glyphosate 251
grana 42, 43
greenhouses, commercial 48–9
greenhouse effect 84
greenhouse gases 84–5
  see also carbon dioxide; methane
grey dunes 34, 35
grey squirrels 17–18
gross productivity 51
ground beetles 60–1
growth of plants 190–5

H
Haber process 90
habitat 3
haemoglobin 126
haemophilia 115–16
hairiness 200
haploid cells 104
Hardy-Weinberg principle 121–3
hatching time 85–6
hCG 212
heart disease 28, 29
heartbeat rate 140–2, 168–9
heat gain centre 202
heat loss centre 202
heat production and loss 197–8
heat stroke 204
heather 51–5
herbicides 70
herbivores 65
herceptin 230–1
heterozygous genotypes 104–5
Himalayan rabbit 102
hippocampus 152
histamine 189
histograms 29, 30
homeostasis 195–209
homozygous genotypes 104–5
horizontal gene transfer 252
hormones 186–90
horses 112
housekeeping genes 227
human activity, succession and 36–7
human adult stem cells 228
human embryonic stem cells 227–8
Human Genome Project (HGP) 101
human populations 20–30
  calculating growth rate 21–4
  demographic transition 24–9
humidity 138
humming bird 65
hyperlipoproteinaemia 104
hyperpolarisation 158, 159
hyphae, fungal 80
hypothalamus 202
hypothermia 196

I
immune response 189
impulses 154–68
in vitro gene cloning 242–3, 247
in vivo gene cloning 243–7
indoleacetic acid (IAA) 192–5
infantile diarrhoea 29
infectious diseases 28, 29
inflammation 189
inheritance 106–18
inhibition 166–7
inhibitory centre 141
inhibitory post-synaptic potential (IPSP) 170
inorganic fertilisers 91–2
insecticides 70–1, 167
  natural 250
insects
  pests 85–8
  succession of in a dead body 31–3
insulin 205–6, 208
integrated pest management (IPM) 74–5
interspecific competition 15–16, 17–18
intraspecific competition 15, 16
introns 218
ion channels, gated 157
islets of Langerhans 205

J
junglefowl 171–2

K
kineses 134–5, 136–8
Krebs cycle 56, 57–9

L
lactate 58, 185
ladybird beetle 72
law of diminishing returns 92–3
Libet, Benjamin 131–2
lichens 13–15
life cycles 87, 213
life expectancy 26, 27, 28–9
life tables 25–7
ligases 244, 245
ligation 244, 245
light-dependent reactions 42, 43, 43–4
light energy 42–7
light-independent reactions 42, 43, 44–7
light intensity 47–8, 54, 55, 83
light-sensitive pigments 146
lightning 90
link reaction 56
liposomes 252, 253
livestock, rearing 65–9
lizards 201–2
locus 99–101
log scales 17, 78
Lotka-Volterra model 18–20
Lund tube 97–8
luteinising hormone (LH) 211–12
lynx 18, 19–20

M
maize 49, 50, 87, 92–3
malaria 127, 253
mammals 65
mantids 65
manual DNA sequencing 259–61
manure, farmyard 91–2
marine fish catch 63
mark-release-recapture 7–8
marker genes 245
marram grass 33–5, 36, 37
mast cells 189
meadow brown butterfly 89
measurements 10
medical screening 253–61
medulla 140–1
meiosis 108
menstrual cycle 210–12
menstruation 210
mesophils 77
messenger RNA (mRNA) 215–16, 232–3
  transcription 217–19
metabolic rate 197, 203–4
methane 84–9
microorganisms 80–1
  breakdown of leaves 78, 79–80
  composting 77
  nitrogen-fixing 90
microsatellites (STRs) 261, 263–4
mitochondria 57–8, 184
mobile dunes 34
monitoring populations 88–9
monohybrid inheritance 106–16
motor neurones 133–4, 153
multiple alleles 112–14
multipotent cells 228
muscles 171–85
  contraction 175, 178–83
  structure 176–8
muscle fibres 176
mutagenic agents 224–5
myelin sheath 154, 161–2
myelinated neurones 161
myofibrils 176, 177–8
myoglobin 184
myosin 177–8
myosin heads 179–80
myxomatosis 38, 119–20

N
NAD 56, 58–9
  reduced 56, 57, 58, 58–9
NADP 43, 44
  reduced 42, 43, 44, 47
naked-neck broilers 67–9
natural insecticides 250
natural selection 120, 123–7
negative feedback 196–7, 202, 206–7
nematode worms 78
nerves 153
  parasympathetic 141
  sympathetic 141, 189
nerve impulses 158, 160–1

Index

rods 145–9

S

saltatory conduction 161
sample size 4–5
sand dunes 33–6
saprobionts 81, 89
sarcolemma 176, 180–1
sarcomeres 177–8
schizophrenia 169
Schwann cells 154
screening, medical 253–61
second messenger model 207–8
secondary consumers 62, 81, 82, 89, 90
selective weedkillers 194
self-ligated plasmids 245
sensitivity 148–9
sensory neurones 133–4, 153
sex, inheritance of 114
sex-linked characters 115–16
shell thickness and strength 70–1
shivering 198
short tandem repeats (STRs) 261, 263–4
shunt vessels 199
sickle-cell anaemia 126–7, 253
sinoatrial node (SAN) 140–1, 168–9
skeletal muscles 174–5
    contraction 180–3
    see also muscles
skin 143–5
sliding filament theory 178–80
slow-flowing streams and rivers 31, 36
slow-twitch fibres 172, 173, 183–5
slugs 60–1
small interfering RNA (siRNA) 231–3
snails 80
snapdragons 110–11
snowshoe hares 18, 19–20
sodium ions 144–5, 155–6, 157, 158–60
sodium-potassium pump 156, 159–60
somatic cells 105, 227
somatic cell nuclear transfer (SCNT) (animal cloning) 214–15
somatic cell therapy 252–3
Southern transfer (Southern blot) 239
sowing density 15–16
spatial summation 165
Spearman's rank correlation test 10, 11, 94–5, 97
specialisation 153
    gene specialisation 226–7
speciation 127–9
species 127
speckled wood butterfly 89
sphincters 198
squids 161

squirrels 17–18
stabilising selection 125–6
standard error 10, 11, 12–13
stands (of heather) 51
starvation of poultry 68
statistical tests 9–13, 69
    carrying out 11–12
    purposes 9–11
stem cells 227–8
sticky ends 236, 244
stimuli 133
    see also responses to stimuli
striated muscle 177
succession 30–7
    and human activity 36–7
sugar cane 55
sulphur dioxide 13–14
summation 149, 165–6
surface area to volume ratio 65
survival 132–4
survival curves 27–9
sweat glands 199
sweating 199–200
sympathetic nerves 141, 189
synapses 133–4, 149, 150, 163–6
synaptic cleft 163–4
synaptic knobs 153, 154, 163, 165
systemic pesticides 70

T

tamoxifen 229, 230
target organs/cells 188
taxes 134, 135, 136–8
temperature
    control of body temperature 195–204
    effect on duckweed 12–13
    global mean temperature 85–8
    rate of photosynthesis 48–9
template DNA 240, 241
temporal summation 165, 166
tendons 174–5
tertiary consumers 62
test statistics 10, 11
testis-determining gene (SRY) 114
testosterone 186–7
thermophils 78
thermoregulation 195–204
threshold value 157, 160
thylakoids 42, 43
tigers 1-2
totipotent cells 226, 227
transcription 216–19
    control of 228–31
transcription factors 228–31
transects 5–6
transfection 252–3

transfer RNA (tRNA) 215–16, 219, 220
transformed bacteria 245–7
transgenic organisms 245, 265
    usefulness of 249–52
translation of mRNA 216, 217, 219–21
    control of 231–3
trends 88
triceps 175
trinucleotide repeats 220–1
triose phosphate 47, 56
tropical crops 55
tropisms 139, 190–2, 193–4
    phototropism 139, 191, 192
tropomyosin 179, 180
troponin 179
tumours 225–6
    breast tumours 229–31
tundra 39
two-way paper chromatography 45
Type I diabetes 208
Type II diabetes 208–9

U

uncontrolled reflex actions 133–4
universality of genetic code 221–3

V

vasoconstriction 198–9
vasodilation 198, 199
vectors 243, 244–5
viruses, plant 87–8
vision 145–9
visual acuity 148–9
vitamin A deficiency 248–9
volcanic vents 59
voltage-gated channels 157, 158

W

water hyacinths 72–4
water vapour 85
weedkillers, selective 194
weeds 69
wheat 49
winter wheat 93
wolves 172
woodland 34, 35
woodlice 79–80, 136–8

X

X chromosome 114

Y

Y chromosome 114
yellow dunes 34